About the Series

IDEAS IN PROGRESS is a commercially published series of working papers dealing with alternatives to industrial society. It is our belief that the ills and profound frustrations which have overtaken man are not merely due to industrial civilization's inadequate planning and faulty execution, but are caused by fundamental errors in our basic thinking about goals. This series is designed to question and rethink the underlying concepts of many of our institutions and to propose alternatives. Unless this is done soon society will undoubtedly create even greater injustice and inequalities than it contains at present. It is to correct this trend that authors are invited to submit short monographs of work in progress of interest not only to their colleagues but also to the general public. The series fosters direct contact between the author and the reader. It provides the author with the opportunity to give wide circulation to his draft while he is still developing an idea. It offers the reader an opportunity to participate critically in shaping this idea before it has taken on a definite form. Future editions of this paper may include the author's revisions and critical reactions from the public. Readers are invited to write directly to the author of the present volume at the following address: Godfrey Boyle, Undercurrents Magazine, 275 Finchley Road, London NW3

THE PUBLISHERS

ABOUT THE AUTHOR

Godfrey Boyle was born in London in 1945, but grew up in Ireland, initially in Belfast and later in Dublin. He studied Engineering at Queen's University, Belfast, but dropped out of university after failing his finals in 1969, largely as the result of his "utter disillusionment with the narrowness and irrelevance of the academic Engineering curriculum". He joined the International Publishing Corporation in 1970 and worked on *Electronics Weekly* for three years as the paper's Communications Correspondent. In 1972, he founded the radical science and technology magazine *Undercurrents* which he edited in his spare time until, in 1974, the magazine became successful enough to employ him as full-time editor. After five years of living in London, he recently moved to a cottage in Gloucestershire, where he intends to "start practising what I have been preaching".

LIVING ON THE SUN

By the same author

Radical Technology: edited by Godfrey Boyle and Peter Harper; Wildwood House, London, and Pantheon Books, New York, 1975.

IDEAS IN PROGRESS

LIVING ON
THE SUN

Harnessing Renewable Energy
for an Equitable Society

Godfrey Boyle

CALDER & BOYARS

LONDON

First published in Great Britain in 1975
by Calder & Boyars Limited
18 Brewer Street London W1R 4AS

© Godfrey Boyle 1975

ISBN 0 7145 1094 7 Cased Edition
ISBN 0 7145 0862 4 Paper Edition

Typeset by Ashley Printing, Liverpool

Printed in England by
Whitstable Litho Ltd., Whitstable, Kent

CONTENTS

to Sally, Holly and my Mother and Father, with love.

INTRODUCTION

This book aims to show that it is entirely possible for the industrial nations of the world to terminate their dependence on non-renewable sources of energy and to create a gentler, fairer, more ecologically conscious civilisation based on the indefinitely-sustainable energies of the Sun, the geothermal heat of the Earth, and the tidal motion of the oceans.

Such a civilisation, as I hope to show, would in no way necessitate a return to the "primitive" conditions which are commonly believed to have prevailed — at least for the poor — in the epochs that preceded our fossil fuel-based industrial age. Renewable energy sources are more than capable of providing enough power to satisfy all the *reasonable* demands of 20th century consumers for domestic heating, lighting, and cooking, for transport, and for manufacturing industry.

The inexhaustible energies inherent in sunlight, in wind and water, in plants, in the earth's geothermal heat, and in the ebb and flow of the tides, provide a far firmer foundation for liberty, equality and fraternity among Mankind than the energy *products* purveyed by the present oligopolistic cartels, both capitalist and state-capitalist.

I am, of course, aware that to advocate such a "back to nature" policy in Britain at the very time when the nation is on the verge of becoming self-sufficient in oil (Scottish Nationalists permitting), may appear to many to be an exercise in futility at best and insanity at worst.

But even the most sanguine of optimists cannot deny that North Sea oil will eventually run out, and that it will be very expensive in monetary terms, in terms of the number of lives lost in drilling in the treacherous conditions that prevail in the North Sea, and in terms of the environmental and social effects on Scotland and its people. Simple prudence — assuming that those in control of Government in this country are aware of the meaning of the concept — would suggest that

9

North Sea oil and gas should be invested in capital equipment for harnessing the *renewable* energy sources, to enable the country to survive, and thrive, in the post-fossil fuel era.

But in the last analysis this book is not particularly concerned to set forth the policies which an advanced technocracy ought to adopt in order to ensure its survival in a world where the cheap, non-renewable energy sources on which it is founded are becoming expensive and increasingly difficult to find.

In the final analysis, indeed, this is not so much a book about energy, or even about technology, as about *political economy,* as Kropotkin[1] defined it: "a science devoted to the study of the needs of men and the means of satisfying them with the least possible waste of energy".

Like Kropotkin, I am far from convinced that "the means now in use for satisfying human needs, under the present system of permanent division of functions and production for profits, are really *economical*". Like the Anarchist Prince, I regard the economic tyrannies that prevail in both state capitalist and capitalist countries as, at best "mere wasteful survivals from a past that was plunged into darkness, ignorance and oppression, and never took into consideration the economical and social value of a human being". And like him I look forward to the creation of a radically new social, political and economic structure "in which men, with the work of their own hands and intelligence, and with the aid of machinery already invented and to be invented, should *themselves* create all imaginable riches".

Such a society, in which men realise that "in order to be rich, they need not take the bread from the mouths of others", cannot be founded on the cynical exploitation of future generations that is inherent in the extensive burning of fossil fuels. It must, as a necessary but not sufficient condition, be woven around the rhythmic but inexhaustible ebb and flow of Nature's renewable energies.

1 Energy and Exploitation

Year in, year out, as it has done for thousands of millions of years, the Sun pours down upon the Earth far more energy than the human race can ever hope to use.

The total amount of solar *power* intercepted by the Earth is about 170 million million kilowatts, which is equivalent, if you like, to 170 million million continuously-burning 1 kW electric fires — or, to use a more personal metaphor, the amount of power that the 3,500 million people of the earth would consume if they *each* left more than 48,000 1 kW fires burning continuously. Added up over a year of 8,760 hours, a *power* of 170 million million kilowatts amounts to no less than one-and-a-half million million million kilowatt-hours of *energy*, because energy is equal to power multiplied by time.

Compared to this enormous annual influx, the amount of energy we humans actually consume in such forms as oil, coal, gas and electricity is extremely small. Our total world consumption of energy per year, in all forms except food, is about 60 million million kilowatt-hours — only one twenty-five thousandth of the annual solar energy input to our planet.

If we use so little of the Sun's energy, where does the rest of it go? For a start, about 30 per cent of it is reflected straight back into space. Another 47 per cent, it has been calculated, goes to heating up the planet's land surface, its atmosphere and its oceans. And some 23 per cent is used to evaporate, from our lakes and oceans, the water which eventually falls as rain and fills our rivers.

These three phenomena, reflection, heating and evaporation, account for *almost* 100 per cent of the earth's energy expenditure, but not quite.

A little of the incoming energy, about 0.2 per cent, is expended when the sun heats different parts of the atmosphere and of the oceans to different temperatures. This causes pressure differences, which in turn cause air and water to flow from areas of high pressure to areas of low pressure. These flows make themselves manifest in the air in the form of *winds*, and in the sea in the form of *currents*. When winds react with the ocean surface, they cause another indirect manifestation of the Sun's energy — *waves*.

An even smaller amount of solar energy, only about 0.02 per cent of the total, is absorbed in the vital process, known as *photosynthesis*, by which plants convert carbon dioxide and water from their surrounding environment into oxygen and carbohydrates (such as starches and sugars), and on which the whole of human, animal and plant life depends.[2] The energy absorbed during photosynthesis is stored in the chemical bonds which hold the plant's carbohydrate molecules together. That energy is released when air-breathing animals use oxygen to break the molecular bonds, which is what happens when a plant is digested as food; or it can be released by cruder methods, such as simple burning, during which process the plant's stored carbohydrate reacts with oxygen at high temperature, and gives off its energy as heat.

When plants die, their leaves are normally decomposed by *aerobic* bacteria in the presence of oxygen, and some of the energy stored in their carbohydrates is released to the environment in the form of heat — which is why a compost heap becomes warm.

But some dead organic matter is deposited at the bottom of lakes or in peat bogs where there is little oxygen, and does not decay completely. When such partially-decayed matter, over millions of years, builds up and becomes buried beneath layers of sand, rock and sediment, it eventually turns into one or other of the "fossil fuels", such as oil, coal, or the tar sands and oil shales that are now being heralded as a solution to the US "energy crisis".

When undergoing partial decomposition in the absence of air (*anaerobic* decomposition), such deposits give off large quantities of "Natural gas". The Natural gas so generated is another fossil fuel, and is composed mainly of methane. Methane can, however, be generated by the anaerobic decomposition of ordinary, non-fossil, organic matter. But more about that later.

12

The fossil fuels represent, in a very real way, the Earth's "life savings" of energy: they amount to the planet's *non-renewable capital* because the geological and biological processes which formed them over the past 600 millions years have proceeded so slowly that the deposits can be regarded as essentially *fixed* in quantity — at least in relation to any human time scale.

By contrast, the energy flowing continuously to our planet from the Sun is *renewable income* energy because it will be available for as long as the Sun keeps shining, in quantities which, though limited, are some 25,000 times as large as the world's total annual energy consumption. Apart from the Sun, there are two other sources of renewable income energy. One is the gravitational pull of the moon and sun, which acts upon the world's oceans and causes tides. The total amount of energy stored in the Earth's tides has been estimated to be about 26 million million kilowatt-hours per year, or about 0.0017 per cent of the total solar energy input. The other is the "Geothermal" energy generated deep in the earth's core by the decay of radioactive substances. The total amount of geothermal energy flowing to the earth's surface, either by direct conduction through the ground or by convection through volcanoes and hot springs, is reckoned to be about 280 million million kilowatt-hours a year.[3]

Throughout history, mankind has found ways of tapping this renewable energy supply — by building sailing ships and windmills to capture the energy of wind, by devising water mills to harness the rain that flows through rivers, by constructing houses with thick walls to store the Sun's direct heat energy, and by burning trees and other plants to trigger the reaction with oxygen which liberates the energy stored in their carbohydrate bonds.

Like the energy stored in the carbohydrates of plants, the energy stored by fossil fuels can also be liberated by reaction with oxygen — which is what we do every time we light a coal fire, start a car engine, switch on the oil-fired central heating, or light a gas stove. But every time we do so, we irreversibly deplete an energy supply that has taken 600 million years to build up. And at present consumption rates, we will have used up the earth's entire supply of fossil fuels in considerably less

13

than 1000 years from start to finish, even taking the most optimistic estimates of the world's recoverable reserves.[4]

The reckless exploitation of our irreplaceable fossil resources that has taken place over the past 100 years or so has been accompanied by exploitation of a different and more personal kind. Man's inhumanity to man has, of course, been a characteristic of all epochs, but the advent of easily-accessible concentrations of energy in the form of coal, and later oil, has amplified many times the power of the exploiter over the exploited, the tyranny of the oppressor over the oppressed.

As Tolstoy observed more than 70 years ago: "If the arrangement of society is bad (as ours is), and if a small number of people have power over the majority and oppress it, every victory over Nature will inevitably serve only to increase that power and that oppression."

The misery of the millions of labourers who worked in the Dark Satanic Mills of 19th century England was made possible (although not *necessary*) by the coal that fuelled the steam engines which drove the primitive machinery of mass production. The alienation of the car workers of Detroit and Dagenham today is made possible by the oil that powers both the sophisticated machinery of mass production and the chromium-plated end-product.

Non-renewable energy resources in general, and fossil fuels in particular, possess unique characteristics which make them eminently suitable for exploitation by ruling oligarchies in the furtherance of their own interests. Fossil fuels are *concentrated* in discrete locations, on land that can be bought (or seized) — unlike the renewable resources, which in general are pretty evenly distributed throughout the globe. Their extraction from the Earth requires large amounts of capital and a high degree of technological expertise, which, as the citizens of the North of Scotland have discovered, means that only Governments and the largest multinational corporations can afford to finance such undertakings. Moreover, since deposits are usually located at a considerable *distance* from the consumers, vast amounts of profit and tax can be added to the cost throughout the various stages of transport and distribution, without the consumer normally being any the wiser. On a gallon of petrol

14

costing 50p in Britain just after the spectacular price increases of October 1973, the price paid to the producer government was about 9.5p, but the markup by oil companies was 11p. Retailers took about 5½p and the Government creamed off a massive 22½p.[5]

Such increases, amounting to more than 500 per cent from producer to consumer, would hardly be tolerated for long if oil energy, like wind energy or water power, could be tapped and controlled by the communities which use it.

So desirable, indeed, have the properties of non-renewable energy sources been in enabling the ruling minorities of society to preserve and extend their influence that when the exhaustion of the Earth's fossil fuel reserves recently appeared as a prospect on the horizon, a search immediately began for substitutes with similarly obliging characteristics.

The energy czars of the world did not have far to look. Their friends in the military establishment had come up with just the answer, in the shape of nuclear fission.

When it comes to shoring up economic and political monopolies, nuclear fission is a well-nigh ideal energy source. Like coal and oil, it depends on digging up something solid and tangible, namely uranium, from ground to which one can own the "rights". What is more, even greater amounts of capital expenditure and technical know-how are required to refine and enrich uranium than are needed to process oil. Better still, the technology involved in turning enriched uranium into usable energy is so esoteric, so costly, and so dangerous that the number of organisations in the world capable of performing the feat is probably only about a dozen at present.

Never mind that the development of nuclear power will bring in its wake an increased death rate from cancer and leukaemia, and a higher incidence of genetic diseases in future generations, even if there are no catastrophic accidents.[6] Never mind that a reactor "melt down" could decimate the population of a city, or that leaks of radioactive waste could poison the world's rivers and oceans. And never mind that plutonium will be used in vast quantities to fuel the highly-unstable breeder reactors needed to make atomic power

15

an economic proposition, and that plutonium is so toxic that just one kilogram of it could, even under optimistic assumptions, cause 2.7 million fatal lung cancer cases.[7] All that matters to the ruling oligarchy is that nuclear power is potentially capable of providing a very large proportion of the energy needed to sustain the status quo for a few more centuries. All right, their plausible propagandists argue, there may be accidents: but if accidents occur individuals are to blame — not the system. And "we" need ever-growing amounts of energy, they will say, because without it growth would cease and international trade collapse. Anyway, they assert, nuclear fission is just a "stop-gap": the scientists will soon succeed in perfecting *fusion* power, which will creat vast amounts of energy from the "virtually-inexhaustible" reserves of deuterium in the world's oceans. Fusion reactors will be the ultimate answer, we are told. What we are not told is why it should be necessary for us to go to immense trouble to create our own fusion reactors when we receive one-and-a-half million million million kilowatt hours of energy a year from the Great Fusion Reactor in the sky known as the Sun. Could it be because no one has yet perfected a way to corner the market in sunshine? It is because no one, so far, has been able to channel all the Sun's energy into a small number of outlets from which it can be divided up into units and sold in packages, as has been done with virtually every other natural commodity?[8]

In so far as energy can be said to have political characteristics, then, the non-renewable energy sources can be classed as hierarchical, authoritarian and exploitative. The Earth's *renewable* energy resources, on the other hand, are non-hierarchical and egalitarian. They are accessible in almost every part of nearly every country to persons possessing a minimum of capital and technical knowledge. And they are non-exploitative not only in the sense that they cannot be "used up", but also in the sense that they are distributed in fairly low concentrations everywhere, so that it is difficult to accumulate inordinate amounts in one place — a fact which is a decided disincentive to the entrepreneurially-minded. For example, the maximum amount of solar power extractible at the surface of the Earth is, at best, somewhat less than one kilowatt per square metre, whereas the amount of power given off by the core of a nuclear reactor can be 65 kilowatts per litre in the case of a Light Water Reactor, or as high as 400 kilowatts per litre in the case of a Fast Breeder Reactor.[9]

Renewable energy sources are, however, unreliable by the standards we have come to expect in a civilisation where everything is "instant". The Sun does not always shine when you want heat, the rivers are not always at their fastest-flowing when you want your water wheel to turn, and the wind, as St John succinctly put it, "bloweth where it listeth".

For this reason, most of the systems that have been devised for tapping renewable energy sources incorporate some degree of storage, so that the energy can be collected when it is available and held until needed. In the case of plants, however, the storage system is inherent: store the plant and you've stored its energy. But large amounts of storage capacity are not so necessary as some maintain. From an ecological point of view, there is considerable merit in basing a civilisation on the rhythm of natural flows. The members of such a civilisation are unlikely to forget that the energies of nature are our life blood and can be over-exploited only at our peril.

In the explorations that follow, I shall concentrate on analysing how the energy requirements of the domestic consumer can be met by the use of renewable sources, partly because proposals advocated for an individual household are relevant in wider spheres, and partly because it is in his home, in his electricity, coal, heating oil and gas bills, that the ordinary citizen feels the impact of energy costs most acutely. But I hope also to touch on the subject of energy use in other major fields, such as manufacturing industry and transport, and to show how such activities could continue, albeit in a radically different and diminished form, in the egalitarian, ecologically-conscious, libertarian, decentralised, low-energy society which we must begin to build — not in the hope that such a step will somehow insure us against the possible collapse of 20th century technocracy, but as a revolutionary act that will help to hasten the collapse of that technocracy.

2 Domestic Energy Consumption

Let us begin by looking at the energy needs of the average British household. Energy consumption figures in the UK are more modest than the hyper-extravagant statistics for the United States, but are far greater than those for the energy-starved third world, so a British example is as representative as we are likely to find. According to the Annual Abstract of Statistics[10] the amount of energy supplied to domestic consumers in Great Britain in 1972 was some 14,389 million Therms. Since one Therm of energy is equivalent to 29.3 kilowatt-hours, and since there are some 18.2 million households in the UK, the total energy consumption per household works out at around 23,000 kilowatt hours (kWh) per year. Any analysis of the amounts of energy consumed in various specific activities is complicated by the fact that each of the various forms of energy — electricity, gas, coal, oil and so on — can be used for several purposes. Electricity is the most flexible energy source, and can be used not only for space heating, lighting, cooking, and water heating but also for powering radio and TV sets and domestic appliances. Gas can also be used to energise all these activities, with the exception of TV, radio and most motor driven appliances. Coal and oil, however, are normally only used for space and water heating. Electricity however, though it may be the most flexible energy source, is generated at power stations in an extremely inefficient manner. For thermodynamic reasons, about two thirds of the energy produced at most power stations is in the form of heat, not electricity. And since in Britain the CEGB has ignored the possibility of using this "waste heat" in so-called District Heating or "total energy" schemes, two thirds of the energy simply vanishes into the air in huge cooling towers, or in raising the temperature of our rivers.

But electricity, on the other hand, is virtually the only energy source capable of supplying power for lighting, radios, TV sets and motor-driven appliances, so I will assume for convenience that electricity is used for these purposes, but not for cooking and space and water heating. Diamant[11] gives 1,200 kWh as the amount of *necessary* electric power consumed by an average household, and that is the figure I will use here.

Assuming that gas is used for cooking, the average domestic cooker in the UK uses about 80 Therms per year[12], a consumption equivalent to about 2,400 kWh a year.

As for domestic *hot* water consumption, Marsh's study in 1971[13] gave a figure of 52 litres per person per day, which, at three people per household on average, works out at around 150 litres per day. To heat 150 litres from the average UK ambient temperature of 10°C to 50°C, the temperature needed for a bath, requires about 7 kWh a day. Since peoples' need for hot water for washing, baths and so on is pretty constant throughout the year, we can get a fair estimate of the annual requirement by simply multiplying by 365, which gives a figure for the total annual domestic hot water energy needs of about 2,500 kWh a year. Adding all these specific requirements up, the total comes to 6,100 kWh—2,500 kWh for hot water heating, 2,400 kWh for cooking, and 1,200 kWh to power lighting and electrical appliances. Which, when deducted from the 23,000 kWh used *in toto,* leaves some 17,000 kWh as the amount of energy consumed on average in space heating. But before attempting to show how these requirements can be met using renewable energy resources, it is interesting to examine whether or not the various amounts of energy are at present being used efficiently, and if not, what reductions in consumption would be made possible by substituting more efficient methods of utilisation.

Space heating is one obvious target for criticism on efficiency grounds. British building regulations, for example, specify that roofs should not lose heat at a rate greater than 1.4 watts per square metre of roof, for every degree of Centigrade temperature difference between the outside and the inside of the house. The corresponding maximum permissible level of heat loss from walls is 1.7 watts per square metre per degree centigrade.

These losses are extremely extravagant by Continental

standards, which are more than twice as stringent.[14] If we were simply to adopt Continental insulation levels, we could therefore reduce the amount of heat lost through walls, roofs and floors to less than half its present level. If we were also to halve the heat loss through the windows of our homes by installing double glazing, and halve the number of ventilating air changes per hour, we could cut the amount of domestic space heating energy consumed in the UK by at least 50 per cent. On average, this would mean a reduction in consumption from 17,000 kWh to less than 8,500 kWh a year per household.

As for the 1,200 kWh of domestic electricity consumption, about three quarters is devoted to supplying power for lighting, which in almost all houses is provided by incandescent bulbs. Just by substituting "warm white" fluorescent lights for the incadescent fittings, the same amount of light can be provided with less than one seventh of the power[15]. Assuming the use of transistorised radio and TV sets, which consume very little power, and sensible but non-extravagant use of washing machines, irons, vacuum cleaners, and similar appliances, it seems very reasonable to suppose that electricity consumption could be sliced to, say, 600 kWh a year per household.

Gas consumption for cooking is another wasteful process at present. As Bell, Boulter Dunlop & Keiller point out[16], "Gas cookers are not designed primarily to conserve energy (put your hand over a boiling pan to confirm this). A grill fully on for 10 minutes might be used to make only a few pieces of toast, but the gas used in this time could supply a full oven on low heat for half an hour." Among the remedies they suggest is better insulation of ovens, coupled with increased use of "pressure cookers, insulated pans, and devices to contain the flame beneath the pan and prevent heat escaping round the sides," together with a return to "the old hay box cooking technique, where food requiring long, slow cooking is heated to the cooking temperature and then placed in a well-insulated container to maintain this temperature for the rest of the cooking time."[17] Further reductions, they propose, could be made by the adoption of quick cooking methods like Chinese stir-frying; the use of a full, well-insulated oven; the sharing of cooking facilities between more people on a communal basis;

and of course by simple eating less cooked food — which, a lot of people maintain, would do us a great deal of good[18]. The amounts of energy saved by these measures would depend to an enormous extent on the user's previous consumption habits and on the thoroughness with which he might adopt energy conserving techniques, but it seems reasonable to assume that the 2,400 kWh a year consumed by the average cooker could be trimmed to about half, or 1,200 kWh a year.

Hot water energy consumption is difficult to criticise on efficiency grounds for most households, since most consumers use electric immersion heaters which have a very high efficiency, (ignoring, for the moment, the huge losses at the power station). The only way of reducing our domestic water heating requirement, therefore, is to use less hot water. And even allowing for the fact that people nowadays probably take baths *more* frequently than necessary, in contrast to previous generations, it seems unlikely that hot water consumption can be reduced to an appreciable degree — except, perhaps, by the widespread adoption of showering instead of bathing, or by the use of shared sauna baths. So for the purposes of this book, I will assume that the domestic hot water energy requirement stays at around 2,500 kWh per household.

After taking all these elementary energy-conserving steps, our average household energy consumption now works out at around 12,800 kWh a year — 8,500 kWh for space heating, 2,500 kWh for water heating, 600 kWh for electricity, and 1,200 kWh for cooking. Even this figure, though only about half the 23,000 kWh used at present, could be improved upon considerably by the use of really effective insulation.

In Scandinavia, houses have been built that are so well insulated that they need no heating input at all — all the energy required even in bitterly-cold Scandinavian winters, is supplied by the body heat of the occupants, supplemented by waste heat from lighting, cooking and similar activities. Such houses have a very efficient ventilation system in which air, though not allowed to go "stale", is changed much less frequently than in most conventional houses[19].

But these steps and others are only really effective in new houses designed especially to take advantage of them. And although all *new* houses from now on should be designed for maximum energy conservation, there are limits to what can be done with existing houses before it becomes more sensible to pull the structure down and build a new one. For this reason, I will stick to the energy consumption figures quoted above, even though these could be reduced very much further by the adoption of additional common-sense conservation measures.

In considering how this reduced, but still substantial, demand for energy could be met by the use of renewable energy sources, I propose to concentrate on those renewable energies which are derived from the Sun's radiation — namely *solar power stored in plants, direct solar power, wind power,* and *water power.* The other forms of renewable energy, tidal and geothermal, are unlikely to have very widespread potential in enabling *individuals* or small communities to supply a substantial proportion of their energy needs, simply because by no means everyone lives on the shores of a sea or lake, and few people live near hot springs or similar natural sources of geothermal heat. The possibility of using these less accessible sources of renewable energy for *industrial* purposes will however be considered in the final chapter.

3 Plant Power

To begin at the beginning, with the Sun: the one-and-a-half million million million kWh of solar energy intercepted by the Earth's plane each year does not all find its way to the surface of the earth. The amount of solar power falling on a square metre just *outside* the earth's atmosphere, as measured from spacecraft, is around 1.4 kW. But the amount of solar power falling on a given square metre down *at* the earth's surface (known as the "insolation") depends greatly on the latitude, the time of day, the time of year, the clarity or otherwise of the atmosphere, and a host of other complicated factors. For the UK, in clear air, Brinkworth[20] gives 900 watts per square metre as the maximum power on a horizontal surface at midday on the summer solstice. At midday on the winter solstice however, the power per square metre falls to under 200 watts.

According to Brinkworth, the total daily insolation *in clear atmospheres* in central UK varies between 8.4 kWh per square metre and 0.8 kWh per square metre, including both direct and diffuse sunlight, and the total annual insolation is about 1,700 kWh. It is an interesting fact that, because of the tilt of the earth's axis, the highest daily amounts of insolation do *not* occur at the Equator, as one might intuitively expect, but at latitudes of around 40°. In mid-summer, even Britain's daily maximum insolation in clear conditions, 8.4 kWh, is greater than at the Equator, 7.5 kWh. (The annual *total* insolation at the Equator is, of course, greater than in Britain: 2,300 kWh, as compared to 1,700 kWh).

But Britain's air, unfortunately, is seldom clear. Due to cloud and air pollution, as Brinkworth points out, the actual annual insolation in the UK is around 900 kWh per square metre — just over half the clear air maximum. Averaged out over night and day for 8,760 hours in a year, this works out at just over 100 watts per square metre.

In many equatorial and tropical countries, however, the air

25

is clear nearly all the time, and the annual insolation is around 2,000 kWh a year — about 200 watts per square metre, on average. Compared to the 1.4 kW per square metre solar radiation level outside the atmosphere, then, the amount received on the earth's surface is, *on average,* between 100 and 200 watts per square metre — say about one tenth.

Long before man thought of tapping this solar energy for himself in the form of heat, or wind, or falling water, he depended absolutely, as he still does, on the energy from the Sun that is trapped by green plants in the process of photosynthesis. As mentioned earlier, photosynthesis provides the energy which enables plants to synthesise carbohydrates and life-giving oxygen from water and carbon dioxide. When we eat these carbohydrates and they react with oxygen, they provide us with the food energy we need to keep our metabolism going, to drive our muscles, to build our tissues, and to power all the other functions of our bodies and brains — either directly, when we eat a plant, or indirectly, when we eat the flesh of a plant-eating animal.

When we burn plant material, we can release this stored carbohydrate energy in the form of heat. The plants which, throughout history, have been burned to release their stored solar energy are, of course, trees. But burning wood from trees in an ordinary fire wastes about three quarters of the stored energy. Up to 85 per cent of the energy can, however, be reclaimed by the use of special stoves which, in addition to burning the wood itself, burn the gases given off by the wood. (In Sweden, during World War II when petrol was scarce, many cars were kept running on "Gengas" from wood burners.)[21] But even if this "wood gas" is not burned, a water boiler installed in the chimney of a stove can boost the efficiency to around 60 per cent. Wood-burning stoves are being used by the Biotechnic Research and Development (BRAD) community in Wales to provide some 35 per cent of the 55,000 kWh of energy needed by the community's 16-person "Eco-house" during the 7-month winter heating season. The timber consumption, about 7 tons per year, is comfortably within the growth capacity of the community's 10-acre hazelwood coppice. But what of the world timber shortage? That shortage, it is true, is due mainly to our enormous "requirements" for packaging and newsprint, and is simply yet another aspect of our profligate over-consumption of the Earth's resources. But it is equally true to say that timber, even if we have enough of it, is far better used in construction, for which it is an almost ideal material, or as a source of valuable chemicals, than as a mere fuel. And in any case what of the city dweller, for whom hazelwood fires are no more than a rustic dream?

One superficially attractive answer might be the burning of household refuse, such as discarded paper, packaging, and wood products. If burned in a suitably-designed incinerator, such refuse would be able to meet at least some of a household's energy requirements.

But it makes no more sense to base one's domestic energy economy on the use of discarded wood products than it does to base it on the burning of wood itself, if we are using up in the process more trees than are being replaced by natural growth — as some countries seem to do at present.

One solution to this dilemma, particularly for urban households, would seem to lie in the cultivation of special *fuel crops*, grown specifically for energy generation rather than food value. Such crops, ideally, would be as compact as possible, and would at the same time contain as large an amount of stored energy as possible in their carbohydrate bonds.

One way of liberating the energy stored in such fuel crops would, of course, be to burn them, as one would do with wood. But there are subtler ways — of which, more in a moment. The first question to be answered is: do such efficient fuel crops exist? And the answer, fortunately, is that they do.

The most important criterion in selecting a good fuel crop is its photosynthetic efficiency — that is, the facility with which it turns sunlight into stored carbohydrate energy.

Most plants in temperate climates convert less than one per cent of the sunshine that falls on them into stored carbohydrate energy. Some plants, however, notably sugar cane, sugar beet and algae, can achieve efficiencies of up to six or eight per cent, when grown under the right conditions.

27

As Brinkworth puts it: "In the most favourable cases, such as the culture of sugar cane in moist tropical regions, where growth can continue all the year round, the annual yield of dry organic matter can be as great as 10 kilograms per square metre of ground surface. Burning this would yield about 60 kWh per square metre for an area where the annual insolation is about 2,000 kWh per square metre. Thus we could recover about three per cent of the Sun's energy in this way."[22]

But as Brinkworth also observes, not only is the supply of carbon dioxide to such quick-growing plants seldom sufficient in a normal, open-air environment, but also, the photosynthetic efficiency of these plants is amenable to considerable improvement: "It is thought that by the breeding of plants specifically for the maximum production of combustible matter and by the most intensive cultivation, yields of the order of 20 kilograms per square metre should be readily realisible *in the short term*. This would represent about six per cent reaping of solar energy."

In this context, Dr D. O. Hall of King's College, London, quotes[23] Taylor and Humpstone's[24] simple but fascinating calculation that if photosynthetic conversion efficiencies of 10 per cent could be achieved (using greenhouses with an enriched carbon dioxide atmosphere) greenhouses covering 700 million acres (only 2 per cent of the Earth's dry land, and four per cent of the planet's cultivated and grazed land) could provide *all the food and fuel* for the estimated 7,000 million world population of the year 2,000. (The calculation, by the way, is based on a mammoth world energy consumption figure of 35,000 million tons of coal equivalent per year, which equals 280 million million kilowatt hours, or more than *four times* our current energy consumption.)

Taylor and Humpstone have also calculated that simply by harvesting and burning five per cent of the annual growth of new plants throughout the world, we could supply as much energy as man consumed in 1970.

For the UK, Dr. Hall has done some surprising arithmetic of his own. Taking the average rate of energy consumption per person to be five kilowatts,[25] including food consumption, he calculates that "if one assumes an average solar radiation of

28

about 125 watts per square metre in the UK, we need about 400 square metres per person at 10 per cent solar energy conversion efficiency. This means that 400 square kilometres (155 square miles) would provide the energy for one million people. London, with eight million people in 620 square miles, would need 1,240 square miles at 10 per cent efficiency, or 20 times its area at only one per cent efficiency."

"Interestingly," he points out, "of this five kilowatts, only about 100 watts equivalent is needed for food to maintain someone using 2,400 kilocalories of food energy per day.

"The UK has a total land area of 94,000 square miles; with a population of 55 million, we would require 8,500 square miles at 10 per cent efficiency to provide *all the energy* for these people in the UK. This only represents about nine per cent of the total land area of the UK."

Though 10 per cent efficiency of photosynthesis is considerably higher than that attained by plants at the moment, Dr. Hall maintains that it is within reach if we do further research on "breeding, genetic manipulation and ecological selection, to produce plants which are inherently more efficient."

Improved techniques, such as growing plants hydroponically in enriched carbon dioxide atmospheres, should also enable higher efficiencies to be achieved by assuring that the plants receive an optimum supply of their essential nutrients at all times.

Taking Brinkworth's figure of only six per cent as the efficiency obtainble in the *short* term, however, means that our average U.K. insolation of 125 watts per square metre would produce an average "plant power" of 7.5 watts per square metre. To supply a power of 5,000 watts per person, therefore, would require some 670 square metres, in theory. In practice, as Brinkworth points out, if the plants were burned in a conventional boiler to raise steam to drive a turbine to generate electricity, the maximum efficiency would be only about one third.

So the actual power generated would only be about 2.5 watts per square metre, and the area needed to supply one person with 5,000 watts would be three times as large, or about 2,000 square metres.

Such calculations are excessively pessimistic, however. It is *not* necessary to assume that only one third of the energy input to a heat engine can be converted into useful energy. If the "waste" heat of the system is regarded as a useful by-product, instead of just a nuisance, then much higher actual efficiencies can be obtained.

Special heat engines — both external and internal combustion engines — have been designed which generate *useful heat* as well as work or electricity. Such systems, called "total energy systems", have become very popular in Sweden in recent years, and are used as the basis of so-called "District Heating Schemes." In these schemes, waste heat — from an engine used to generate electricity for a small area — is "piped out" to local households. Efficiencies of 80 per cent and more are attainable in this way, compared to the 33 per cent typically obtainable if the engine is used to generate electricity alone.[26]

In the example above, the use of an 80 per cent efficient heat engine would give a power of 6 watts per square metre of plant area, and would mean that 5,000 watts of power per person could be generated using only about 830 square metres of plant-growing area. For a population of 55 million, the area required to supply all their present energy needs would be about 45,000 million square metres — about one third of the UK's arable land area, and about one sixth of its total land area. This compares not unfavourably with Dr. Hall's hypothetical 10 per cent efficient photosynthesis system. The 830 square metres needed to supply 5,000 watts of continuous power per person, although just over twice the land area which Dr. Hall's calculations require, still only implies a need for about 18 per cent of the nation's land area (roughly 36 per cent of our arable land) to grow crops that would supply all our *present, extravagant* energy needs, and provide 2,400 kilocalories a day of food on top.

The beauty of such a system, for those of us who are in general opposed to the large scale, centralised production of energy — or anything else — is that it would be very difficult to "scale up". To distribute the heat generated by a total energy system over great distances and large numbers of people (though theoretically possible given well-enough insulated pipes or "heat pipe" techniques) is a far more complicated and cumbersome task than the long-distance transmission of electricity using flexible wires. So plant-powered total energy systems would only be practical on a fairly small scale, or at the individual household level.

Household energy consumption, as we've seen, can reasonably be set at around 12,700 kWh a year — equivalent to a continuous power of some 1,450 watts. To meet this *in total* using plant power would require a growing area of 240 square metres — say a plot of 20 yards by 15 yards, or less than one-sixteenth of an acre. Spaces of this size are available in many medium-sized back-gardens.

But to attempt to service the entire energy needs of a household, let alone those of an entire country, *solely* by burning high-efficiency plants is not a particularly good idea when there are many other renewable energy sources — direct sunlight, wind energy, and water power, for example — capable of meeting those needs either partially or fully.

The balance of all these renewable energy sources varies enormously from situation to situation, and any harmonious solution to a particular energy supply problem will involve a careful assessment of the conditions that prevail at a given site. Such a solution will involve the use of plant power in some situations, and not in others, depending on whether the site has strong winds, a nearby stream, a lot of sunshine, adequate rainfall, good plant-growing conditions, and many other factors.

There are more ways of tapping a plant's hidden energy than simple burning, however. A gentler method, much more closely modelled on natural processes, is to turn most of the stored energy into the form of *methane* by *anaerobic decomposition*.

Organic matter, when it dies, can decompose in one of two ways: *aerobically*, or *anaerobically*.

Aerobic decomposition, as mentioned earlier, is the process which occurs in a garden compost heap. Air-breathing, or *aerobic* bacteria break the carbohydrates down into simpler constituents, which can be used as fertiliser. In the

31

process, a considerable amount of heat is generated. *Anaerobic* decomposition, however, is a process which takes place at the bottom of stagnant ponds, in marshes, and anywhere else where there is no oxygen, and is caused by non-air breathing, *anaerobic* bacteria. During anaerobic decomposition, considerable quantities of methane gas are generated, and the nitrogen-rich residue, as in the case of aerobic decomposition, makes excellent compost. In nature, the methane generated during anaerobic decomposition "bubbles" up to the surface and simply vanishes into thin air. But if the decomposing organic matter is sealed instead in a tank (along with a few anaerobic bacteria to start the ball rolling), and kept at a temperature between 85 and 140 degrees F, the methane given off will be trapped in the tank and can be piped off for use, say, in cooking, or in powering engines. What's more, the residue or "slurry" that's left over can be used as fertiliser. As Lawrence D. Hills of the Henry Doubleday Research Association puts it: "Methane is the most efficient power source we have, for it allows us to 'eat' our compost and have it too."[27]

The precise microbiological processes upon which anaerobic digestion depends are extremely complicated, and the amount of methane one can expect to extract from a given weight of organic input to a "methane digester" varies a great deal. In theory, the input should consist of carbonaceous and nitrogenous matter in the ratio 30 to one, but in practice, the ratio varies because the "digestibility" of the carbon and nitrogen differs from one class of organic matter to another.

Of the high-carbon-content compounds, sugars are the most "digestible" because their molecules are short and easy for the bacteria to break up. As for compounds rich in nitrogen, the best source is usually animal and human excreta. "Urine," as Colin Moorcroft points out[28], "is an exceptionally good source of nitrogen (producing five to ten times as much daily as faeces) but is the most variable, as it, in effect, takes up the slack of the body's nitrogen metabolism." Other promising possible sources of nitrogen for anaerobic digestion include the blue-green alga *Anabena*, and comfrey.

If one were only to decompose the toilet wastes of an average

32

household in an anaerobic digester, the amount of gas produced would only be about 3.6 cubic feet per day, according to Bell, Boulter, Dunlop and Keiller. By adding organic wastes (like tea leaves and vegetable scraps), they say, we could expect to raise the daily gas output to some 10 cubic feet — a lot less than the 32 cubic feet consumed each day by the average gas cooker. Even assuming that this consumption might be halved by the efficient use of burners and so on, that still leaves a considerable shortfall to be made up if methane is to be a serious candidate to fill this important household energy need. Another possible source of organic matter for digestion, which has been experimented with by Dr. S. Klein of the University of California, is the paper, cardboard and kitchen waste fraction of dustbin refuse, pulverised and mixed with water. Lawrence Hills claims that "every pound of this material will produce 13 cubic feet of methane", and points out that "in 1968 every family in Britain threw away 16 lbs every week".[29]

But perhaps the most fascinating approach to boosting methane production is that suggested by Colin Moorcraft.[30] As he points out, it is essential to feed the anaerobic digestion process with the appropriate materials: "Human excreta, especially urine, forms a fine nitrogen source but needs a large supply of carbon to balance it". Kitchen wastes do not constitute such a supply, he maintains, because in the first place it is wrong to put a premium on high per capita production of food wastes; secondly, because kitchen waste is in any case unlikely to contain much carbon since it consists of the tough, fibrous, hard-to-digest stuff that gets thrown out; and thirdly, because "there's a place for the less digestible wastes to go — the compost heap and thence to the soil where the remains of the fibres will improve the soil's structure."

To supply the large amounts of carbon needed, "fodder crops, especially fodder radish, and *sugar beet* most probably constitute a more suitable source," Moorcraft believes. "Furthermore, they are storable (that's what they are designed for) so they can be fed to the digester as and when required."

Having proposed such a system, says Moorcraft, "The next logical step was to consider where the fodder crops might come from. The obvious answer was to grow them using the output of the digester." Conveniently, "most of the nutrients in the output of such a system should be in the liquid fraction, and should therefore make a very convenient pipable feed for soil or sand culture." Having established one "cycle" — fodder crops and human wastes feeding a digester, to generate both methane and a nutrient slurry which is then piped back to fertilise the fodder crop — it then became easy to set up another, equally useful, "cycle" to utilise those nutrients generated by the system in *excess* of the amount needed to fertilise the fodder crop. These additional nutrients — consisting of some of the tops of the fodder crops (although most of these would be fed to the digester), some or all of the solids separated from the output "slurry", and a small proportion of the liquid fraction of the slurry — can be used to fertilise crops grown to feed the people using the digester. Moorcraft suggests that these crops, as well as the fodder crops for methane production, could be grown by the "Sharder" process of soilless, "organic hydroponic" culture advocated by James Sholto Douglas.[31]

"In India," says Sholto Douglas, "where household hydroponics cultivation spreading amongst village and urban societies, it is considered that the simple method of hydroponics can supply eighteen hundred persons with a good meal of three pounds of green food daily throughout the year, from each acre of soilless garden under cultivation. This equals a yield of over eight hundred tons annually per acre."

Moorcraft and his fellow Street Farmers are now testing the potential of this 'bi-cycle' (food loop plus fuel loop) system in their "street Farmhouse II", an ordinary urban house in Crystal Palace, London, which they are hoping to "re-service . . . in ways amenable to direct control by its users."

The New Alchemy Institute, on Cape Cod, Massachussetts, USA, is also looking at a number of ways in which the generation of methane can be integrated with food production.[32]

One, which they call "Sludge Hydroponics", is similar to Moorcraft's proposal, and the Institute lists its advantages in these terms: "Plants grown hydroponically in sludge —

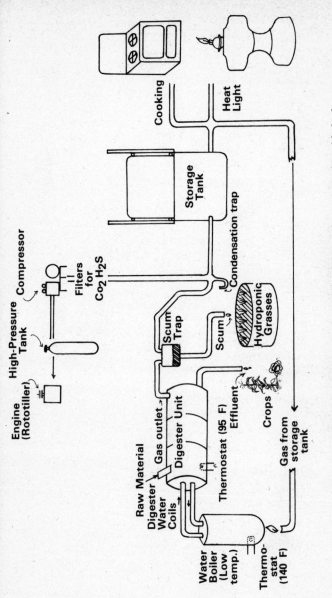

New Alchemy Institute's system for integrating methane generation with food production.

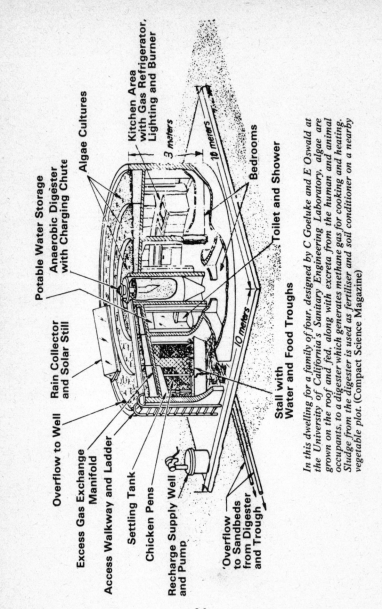

Potable Water Storage

Anaerobic Digester with Charging Chute

Algae Cultures

Kitchen Area with Gas Refrigerator, Lighting and Burner

3 meters

10 meters

Bedrooms

Rain Collector and Solar Still

Overflow to Well

Toilet and Shower

Excess Gas Exchange Manifold

Access Walkway and Ladder

Settling Tank

Chicken Pens

Recharge Supply Well and Pump

10 meters

Stall with Water and Food Troughs

Overflow to Sandbeds from Digester and Trough

In this dwelling for a family of four, designed by C Goeluke and E Oswald at the University of California's Sanitary Engineering Laboratory, algae are grown on the roof and fed, along with excreta from the human and animal occupants, to a digester which generates methane gas for cooking and heating. Sludge from the digester is used as fertiliser and soil conditioner on a nearby vegetable plot. (Compact Science Magazine)

36

enriched solutions can serve a variety of purposes . . . (1) They can do away with the cost and energy of transporting liquid fertiliser to crop lands since they can be grown conveniently near to digesters. (2) They tend to be more productive than conventional soil crops, and thus can serve as a high-yield source of fodder, compost, mulch or silage. (3) They can serve as convenient high-yield sources of raw materials for the digester itself." (See illustration on page 35)

Another New Alchemy Institute proposal is the "Sludge-Algae-Fish" or "aquaculture" system, which "consists of placing sludge into ponds and stimulating the growth of algae. The algae are then used as feed for small invertebrates or fish growing in the pond. The idea is modelled after oriental aquaculture systems."

A third system, (see illustration on page 36) based on the work of Goleuke and Oswald at the Sanitary Engineering Research Laboratory (SERL) in Berkeley California, is the "Sludge-Algae-Methane" process, in which "green algae is grown in diluted sludge, then harvested, dried and digested to produce methane for power and sludge for recycling." Such a system has been experimented with by the Street Farmers[33] in Street Farmhouse One, but the difficulties with algae growing which have been encountered have led them to explore the use of fodder crops instead, along the lines of Moorcraft's suggestions.

The importance of such developments can hardly be over-estimated. If successful, they will contribute to a genuine, sustainable solution to two of mankind's greatest problems — the problem of how to generate enough energy without exhausting or poisoning the earth, and the problem of how to feed our population in a manner that is simultaneously more productive and more ecologically-sane than the proposals of the fertiliser and pesticide companies, the monocultural maniacs, and the "green revolutionaries."

As L. John Fry puts it in the prologue to his description of a simple, low cost, methane generator: "To those on the land eking out an existance, I dedicate this unit. As a morsel of technology, it might well benefit them more than a man standing on the Moon."[34]

4 Direct Solar Power

Though the use of plants may constitute the simplest and most convenient means of tapping the Sun's energy, (since plants are low in cost and have an inherent energy storage capability) there are, of course, *direct* ways of harnessing the Sun's power.

The most direct of these is simply to allow the heat of the sun to warm things up. Frequently, no special sun-catching gadgetry is required — as can be seen by examining the design of buildings such as St. George's school in Wallasey, (see illustration on page 40) where large, south-facing, double-glazed glass walls allow the building to trap as much solar energy as possible, and the school's use of materials with high heat capacity enables it to store captured heat for use during prolonged periods of overcast weather.[35]

Another ingenious but simple method of capturing the sun's heat is the Trombe-Michel solar wall. (See illustration on page 41). Installed at Prof Trombe's famous solar laboratory in the Pyrenees, the wall consists of a large vertical block of concrete which acts both as a structural wall of the house and as a storage medium for solar heat. It faces south and is covered by a double layer of glass, with an air space between the glass and the concrete. As sun shines through the glass, its heat is captured by the concrete surface, which is painted black to increase its absorbency. In accordance with the "greenhouse effect", heat re-radiated from the black wall cannot penetrate the glass, because it is of a longer wavelength than the incident heat energy from the sun, and the double glass lets through only the shorter wavelengths. So the re-radiated heat warms the air in the space between the concrete and the glass. Trombe and Michel have taken advantage of the natural tendency of warm air to rise by incorporating air ducts at the top and bottom of the concrete wall. These allow cold air from the floor of the house to enter the space between the glass and the blackened concrete, where it is warmed, rises, passes out of the ducts at the top and back into the house, so heating it. This elegant system of natural air convection uses absolutely no moving parts. Figures of performance from prototypes tested in the Pyrenees suggest that one square metre of such "glazed walling" can heat 10 cubic metres of well-insulated house, and can provide between two thirds and three quarters of its winter heating requirement. A major reason for this efficiency is the fact that vertical south-facing walls in middle or higher latitudes receive almost four times as much solar radiation in mid-winter on clear days as they do in mid-summer (because the sun is "lower" in winter). A wall 35 cm thick can apparently store about half the heat it absorbs.[36]

A somewhat simpler technique is used by the American designer Steve Baer.[37] Baer's house in the New Mexico desert has its south-facing walls made up of large sheets of double glazing, as in the Trombe-Michel house. But instead of using blackened concrete blocks to store the sun's heat, Baer uses black painted oil drums (there's a certain irony here)

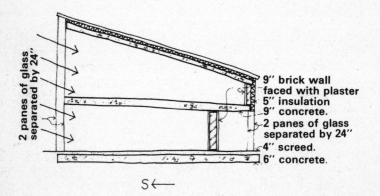

This cross section of St George's School, Wallasey, Cheshire, shows how the building's South-facing glass wall traps as much incident solar energy as possible, while its massive, highly-insulated structure retains the trapped heat for long periods.(Architectural Design)

40

stacked on their sides in a wooden matrix. The drums are filled with water, which has a very high heat capacity, and so absorbs a great deal of the solar heat during the day. At night, the heat stored by the glazed oil drums is 'shut in' by means of large "doors" of highly-insulating styrofoam, covered with a reflecting aluminium skin, which prevents the drums from re-radiating their heat outside, (re-radiation is, of course, partially prevented by the double glazing, but obviously not completely.) At night therefore, the only place the drums can radiate their heat is inside the house. The actual flow of heat from the drums can be regulated by heavy curtains, just in case the drums radiate too much of their stored energy for comfort.

In another sunny area, Phoenix, Arizona, a different method of using water for solar heat storage has been developed by Harold Hay.

Hay's scheme[38] employs plastic bags filled with water and positioned on top of the blackened surface of the flat roof of the house. Above the bags, insulating panels can be moved either to cover or expose the bags. During winter, when heating is required, the bags are left uncovered in daytime. When

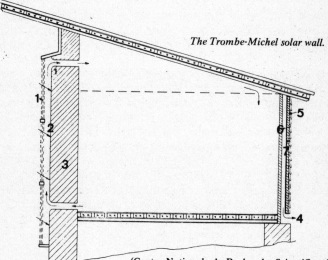

The Trombe-Michel solar wall.

(Centre Nationale de Recherche Scientifique)

41

winter sunshine shines on the blackened surface on which the bags rest, it warms the water in the bags. Then, at night, the bags are covered by the movable insulation panels, which prevent their heat radiating away to the night sky. Instead, the bags radiate their heat through the blackened ceiling into the house. In this way, Hay claims, radiated heat from the ceiling kept the room temperature in his prototype house between $66°$ and $73°$ F, even when outside temperatures fell to freezing point.

By reversing the process — keeping the bags covered by the panels during hot summer days, and letting them cool to near freezing point during clear summer nights by removing the panels — Hay has also been able to keep his house at similarly comfortable temperatures even in the peak of summer heat.

The use of solar energy for space heating, therefore, can be accomplished without the need for fancy equipment, such as that employed in the classic 'Thomason' and 'Dover' houses. But these more complicated systems may be useful if it is desired to convert *existing* houses to use solar energy.

In the Thomason house[39], an existing roof is covered with insulating material, then by blackened sheets of corrugated aluminium, and finally by one layer of transparent polyester film followed by one layer of glass. A perforated pipe running along the apex of the roof allows water to trickle across the black corrugated aluminium surface. Since the surface is heated by the sun's rays ("trapped" by the greenhouse effect) the water trickling over it also becomes warmed. The warm water is then collected by a trough at the bottom of the roof, and piped to a large 1,600 gallon hot water tank surrounded by a "bin" containing 50 tons of fist-sized rocks, occupying a volume of 1,750 cubic feet. The rocks, which have a very large heat capacity, are able to store the heat generated by the hot water tank, and to keep the bin warm during periods when there is no sun and therefore no inflow of warm water. (During cloudy periods, the flow of water from the roof is turned off to avoid lowering the temperature in the tank by the addition of cold water: this is standard practice in many similar solar heating systems).

The heat stored in the rocks is then transferred to the house by air blown through the rock bins and into ducts leading to all the rooms.

The BRAD Community in Wales recently completed a solar roof rather similar in design to Thomason's. BRAD's roof differs from Thomason's in that it uses "Mitlite" glass reinforced plastic sheets instead of polyester and glass. The roof is used for providing hot water, not for space heating. The water flow is turned off during cloudy periods by a simple electronic temperature-measuring system. Details of the roof's construction and cost are shown on pages 44 and 45.

The Dover house, designed by Dr Maria Telkes, differs from Thomason's design in that it uses *air* circulating over the blackened metal sheets instead of water, and in that the storage medium is not rocks but a heat-storing chemical called Glauber's salt (sodium sulphate decahydrate, $Na_2SO_410H_2O$). Glauber's salt is capable of storing eight times more heat than water of the same volume between the temperatures 77° and 98° F. But of course it costs more than bins of virtually free rock, even though it affords a more compact means of heat storage. In the Dover house, the Glauber's salt is stored in sealed five-gallon cans having a volume of 470 cubic feet, over which air is passed (as in the Thomason house) to be warmed before being circulated for heating purposes to the house.

Another even more efficient method for storing heat has been developed by Dr. J. Schroeder and his colleagues at the Philips Research Laboratories in Aachen, West Germany (see illustration on page 46). In the Philips system, heat is stored in so-called "eutectic" (having a low freezing point) mixtures of metallic fluorides. The heat can be released from the storage medium at a precisely controlled rate by means of a partially evacuated "jacket" surrounding the store, the thermal conductivity of which can be varied by varying the amount of hydrogen gas allowed to enter the jacket. The Philips unit, however, is viewed by the company (at least initially) merely as a more efficient replacement for the familiar "night storage heater", and uses electricity as its energy source, rather than hot air or water — though there is no reason in principle why the system could not be adapted to use such sources.

Another, less exotic, heat storage medium is humble

paraffin wax which, as Brinkworth observes,[40], "melts at 55° C with a latent heat of fusion of 40 watt hours per kilogram. This energy is given up again at this rather convenient temperature upon cooling. A storage of 150 kilowatt hours by this means requires only some four cubic metres of space".

The solar collectors installed in the Thomason and Dover houses are, basically, very large versions of what has come to be known as the "flat-plate collector."

Such collectors, as well as being useful for space heating, are very popular in sunny countries like Israel, South Africa and Australia, where they are used for hot water heating. A typical solar hot water heater consists of a blackened metal plate, covered by a transparent layer of glass or plastic, which absorbs the sun's radiation. The absorbed heat is then transferred from the surface either by trickling water over it, or by passing the water through pipes embedded in the plate. The heated water is then stored in an insulated tank and drawn off as needed. (See illustrations on page 47).

A simple flat-plate water heater can be made by adapting a panel-type central heating radiator. One side of the radiator is

BRAD's solar roof is built on top of a large wooden extension to the community's farm house. The diagram overleaf shows the plumbing system employed, and the table shows the cost of the system, excluding labour, at mid-1974 prices. (Photograph by Philip Brachi)

44

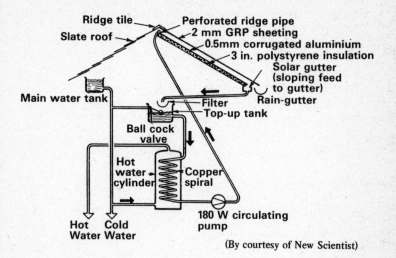

(By courtesy of New Scientist)

BRAD's solar roof: what it cost

Component:	Cost (£)	Cost/sq.m (£)
Corrugated aluminium	132	2·20
Glazing bars + fittings	50	0·83
Plastic sheeting (GRP)	246	4·10
Copper ridge-pipe	18	
Top-up tank	4	
Extra plumbing	10	
Black box control system	5	
Solar gutter	8	
Pump	17	
Fastenings, etc	4	
Total cost	**494**	**8·23**

painted matt black to increase its absorbency, while the other is covered with a piece of insulating material to minimise the amount of heat re-radiated from the back. The black painted side is then covered with a layer, or two, of glass or plastic sheeting, and pipes are connected to the inflow and outflow points on the radiator, to carry the water to and from the hot water tank.

A method of hot water heating that is cheaper and simpler even than the use of modified central heating radiators employs transparent polythene bags, filled with water, with one side painted black. The bags are then left out in the sun. Sunshine falling on the black under-surface warms it, and the surface in turn warms the water.

The efficiency of flat-plate collectors, and most other types of solar collector, can be improved considerably by the use of special "selective absorber" coatings instead of matt black paint. Selective absorbers, which consist of a thin layer of metal oxide on top of a polished surface, are more effective than black paint in enabling the collector to absorb short

PHILIPS THERMAL ENERGY
STORAGE CONCEPT

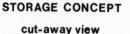

cut-away view

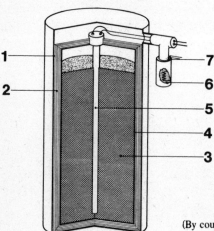

1 Outer vessel

2 Radiant barriers

3 Heat storage media

4 Inner vessel

5 Heating element

6 Getter pump

7 Element for getter pump

(By courtesy of Technology Ireland)

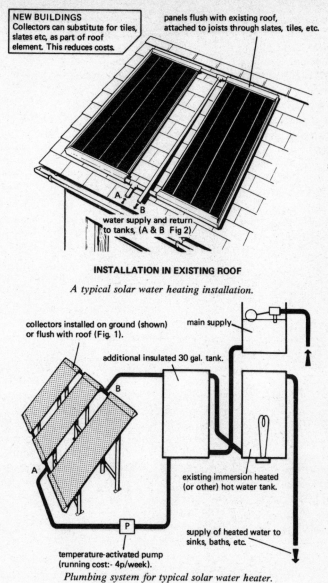

NEW BUILDINGS
Collectors can substitute for tiles, slates etc. as part of roof element. This reduces costs.

panels flush with existing roof, attached to joists through slates, tiles, etc.

A

B

water supply and return to tanks, (A & B Fig 2)

INSTALLATION IN EXISTING ROOF

A typical solar water heating installation.

collectors installed on ground (shown) or flush with roof (Fig. 1).

main supply

additional insulated 30 gal. tank.

B

A

existing immersion heated (or other) hot water tank.

P

supply of heated water to sinks, baths, etc.

temperature-activated pump (running cost:- 4p/week).

Plumbing system for typical solar water heater.

(By courtesy of Conservation, Tools and Technology Ltd)

wavelengths while minimising the re-radiation of longer wavelengths. But such coatings are very expensive, at least at the moment.

The amount of solar energy that can be absorbed by a flat plate collector depends, obviously, on whether or not it is directly facing the sun.

For solar collecting panels which are not attached to a roof, of course, the angle of tilt can be varied at will to suit the variation in the angle of the sun's height at noon from winter to summer. The mean angle of the noonday sun to the vertical in a particular location is equal to the latitude of the location, so it would seem obvious to make the angle of tilt of the collector perpendicular to the mean solar angle — or, in other words, to make the angle of the collector to the horizontal equal to the angle of latitude. In practice, to boost the input of heat during winter months, the collector is often tilted to the horizontal at an angle equal to the latitude plus fifteen degrees.

The vertical orientation of the Trombe-Michel solar wall means that it picks up heat very efficiently during the winter because the sun never rises very high in the sky, but does not pick up so well in the summer — when its heat, of course, is not needed anyway. But for panels which are *built into* a roof, a thorny problem arises, as Robert Vale points out in his Solar Collector Study[41]. The problem is that if a solar collector is to be an integral part of the roof, and if the latitude of the house is such that a steep angle of collector is necessary to ensure the maximum radiation per square metre, then the roof also has to be sloped at a very acute angle, which means that a lot of the space inside the house is wasted. And if the roof is made asymetrical, with one small, sharply-sloping surface, then the collector and another large, gently-sloping surface, then the collector is not only reduced in potential area, but the gently-sloping surface, being north facing and of large area, allows a good deal of heat to be lost.

Vale set out to discover whether it was better to have a small collector set at the theoretically correct angle, or to have a larger collecting area sloped at a less than optimum angle. His studies show that it is better, in a house in which the solar collector is integral with the roof, to have as large an area of collector as possible, even if its slope is less than optimum, because the increased collecting area more than makes up for the decreased power input per square metre.

But especially in an unpredictable climate like Britain's, the specific amount of energy obtainable from any kind of solar collector is very difficult to predict — so much so that computer programmes have been developed to try to take the tedium out of some of the calculations. There are a number of ways of "guesstimating" what to expect under certain conditions, but none is very satisfactory, and the only way to be accurate is to instal instruments at the site in question to measure over a long period the *actual* amounts of direct and indirect sunlight, the frequency and intensity of cloud cover, and so on.

In Britain in winter, the use of flat plate collectors to heat water, either for washing or for space heating, is of very limited effectiveness, because of the increased amount of heat lost by the collectors to their environment in cold weather. The amount of heat lost by a body, according to Stefan's law, is proportional to the fourth power of the temperature difference between the body and its surroundings.

But Andrew MacKillop estimates[42] that a large, 35 square metre array of flat plate collectors — assuming, perhaps optimistically, "a winter ambient solar energy of 150 watts per square metre for 10 hours a day" — should generate some 12 kilowatt hours of heat a day if one accepts a rather low water temperature of only 17.5°C. This temperature would have to be boosted, at least for washing purposes, by some auxiliary heating system, such as a wood-burning stove or an immersion heater, but 12 kilowatt hours a day, over a heating season of, say, 30 weeks, adds up to over 2,500 kilowatt hours a year — nearly a third of our 8,500 kWh space heating requirement.

Hot water for washing, unlike space heating, is something which a household needs all the year round. And although in winter as we've seen, flat plate collectors need to be very large in order to make a significant contribution to a home's requirements, the same is not true of the summer.

In summer, MacKillop has calculated that each square

metre of flat plate collector can extract about 1.32 kWh per day in the period April to September, with an average radiation period of 12 hours, an ambient radiation level of 500 watts per square metre and a water temperature of 45°C.

Assuming an average household in the UK consumes some 220 litres of hot water per day (rather more than Marsh's figure would suggest), which would require some 8.5 kWh to heat from supply temperature to 45°C, MacKillop points out that "an array of six square metres in area should easily supply this amount of heat in most areas of the UK south of Yorkshire-Lancashire . . . provided the background constraints are satisfied." In more northerly regions, slightly larger arrays would be required, but "in mid-summer," he maintains, "in most areas of Britain the high levels of solar energy — up to 850 watts per square metre — will ensure that this heating demand is more than satisfied." Flat plate collectors, then, have considerable potential, both as a source of hot water during the summer, and as sources of low temperature heat during winter. Low temperature heat can be of considerable importance, as we shall see later.

The advantage of the flat plate collector — that it can pick up the sun's energy in the form of diffuse sunshine, as well as receiving direct sunlight as long as it is pointed in the right general direction — is also its disadvantage. Because the solar energy is captured over a wide area, the temperature to which the collector surface can rise is limited, since as the temperature of the surface due to solar heating rises, so does the rate of heat loss of the surface to its surroundings. When the heat loss rises to a level equal to the heat input from the sun, then the temperature of the surface cannot rise any further. But by *concentrating* the Sun's rays, using a lens or a mirror, onto a small surface (which because of its small size cannot lose heat as quickly as a large area) the temperature of the surface can be raised to a very high level. Collectors which concentrate the Sun's rays are called *focussing collectors*. Their disadvantage, of course, is that they have to be pointed directly at the sun, and moved as the sun moves across the sky. Various types of focussing collectors have, however, been developed which only need to be moved infrequently — say every month. These collectors can be thought of as a compromise between flat plate and focussing collectors, and combine some of the advantages and some of the disadvantages of both. (See illustration on page 52).

All the solar energy collecting systems considered so far have used either the direct rays of the sun, or, to some extent, the diffuse sunlight that is present during the day even when the sun is obscured by cloud or haze.

But even when the earth seems cold and dark, its temperature is still very much greater than it would be if the sun were to stop shining. The deepest Arctic winters, where temperatures reach 50 degrees below zero, are still very much warmer than the absolute zero of -273°C that our planet's temperature would sink to without any external source of heat. The earth, therefore, contains in its air, in its rocks, in its soil, it rivers and its seas, a vast amount of stored solar energy — enough to keep the planet's temperature well above zero centigrade in most parts of the world, and enough to prevent temperatures from dropping to anywhere near *absolute* zero, even in the most inhospitable regions.

One ingenious way in which this vast store of hidden energy can be tapped was suggested by Lord Kelvin in 1852, in a paper entitled "The Power Required for the Thermodynamic Heating of Buildings". Kelvin's suggestion has evolved into a device known nowadays as a *heat pump*.

A heat pump works, essentially, like a domestic refrigerator. In a referigerator, the heat energy of the stored food is absorbed by cooling fins inside the cold compartment, and is pumped to another set of fins or tubes outside the compartment, which release the absorbed heat to the environment. This released heat can be felt by anyone who puts his hand into the back of the average kitchen refrigerator. If one were to separate, physically, the heat-releasing tubes from the back of the refrigerator, but still leave them connected (say, by flexible pipes) to the pumping mechanism, and move the refrigerator "box" outside one's house and into the garden, then one would have a crude form of heat pump. One would be receiving heat inside the house, from the heat-releasing fins, as a result of a cooling process taking place

51

outside the house — in the air. The "refrigerator" would now be heating your house by cooling down the air in your back garden. (See illustration at top of page 53).

The energy required to "pump" the low temperature heat from your garden up to the relatively high temperature needed to heat your house is, of course, supplied by the electric motor which drives the refrigerator. But the interesting, and significant, thing is that one can actually get more heat energy out of the system (in the form of heat in one's home) than one puts in (in the form of electrical energy to the motor). The difference, needless to say, is the amount of heat abstracted from the air outside the house. The ratio of the amount of heat energy given out by a heat pump system to the amount of energy used to drive the pump is called the "co-efficient of performance" (COP) of the system. Typical heat pumps have a COP value of about 3:1 — which means that they give out three units of heat for every one unit of work or electricity put in.

The heat absorbed by a heat pump must, of course, come from *somewhere*. The air in the back garden is one source, but

PARABOLIC MIRROR CYLINDER

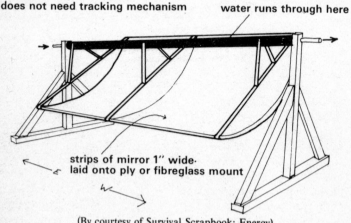

does not need tracking mechanism   water runs through here

strips of mirror 1″ wide·
laid onto ply or fibreglass mount

(By courtesy of Survival Scrapbook: Energy)

The Heat Pump Cycle

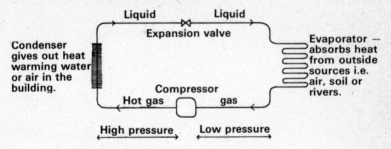

NOTE: Action may be reversed and heat extracted from building in summer.

Schematic diagram of a typical heat pump.

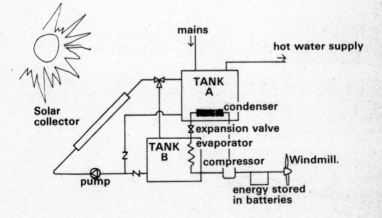

A heat pump can be employed in Winter to raise the temperature of water from solar collectors to a useful level. The heat pump compressor could be powered by a wind generator, as shown.

(By courtesy of Pete Stellon: Architectural Association)

perhaps not the best choice, since air does not have a great capacity to store heat. This means that in order for air to transfer any appreciable quantity of heat, one must pass a very considerable volume of it over the refrigerant vapour, which in turn necessitates the use of a powerful fan to shift the air in sufficient quantity. But air-medium heat pumps of this kind are used extensively in the United States, because they can act as air conditioners during the summer. Another heat source with better heat capacity is the soil beneath one's garden — but to use it necessitates digging a lot of fairly deep holes to bury the pipes that carry the refrigerant vapour. An even better source, if it is available, is the water in a nearby stream or lake. Water has a very high heat-retaining capacity, so that only a small volume of water needs to pass over the refrigerant vapour in order to transfer an appreciable quantity of heat to it.

Another important source of heat could well be an array of flat plate collectors which, as we have seen, can generate considerable quantities of low-temperature heat even in winter. A heat pump is the ideal device for raising this difficult-to-use winter solar energy to a more useful temperature. (See illustration at bottom of page 53).

There are, naturally, a few practical limits to the amount of heat that can be extracted by heat pumping methods.

If too much heat is extracted from a given amount of water, it will, of course, freeze. Heat pumps using air as their medium can also freeze up if the air contains appreciable quantities of water vapour. The same comment applies to soil-source pumps unless the soil is very dry.

But freezing is not a significant limitation if it is anticipated and a sufficient flow of the medium is provided to ensure that its temperature never falls to freezing point.

The BRAD community in Wales is aiming to use a heat pump that cools a tiny stream by one or two degrees C and provides some 12 to 14 kilowatts of power, for an electricity input of only 3 to 4 kW. Even in winter, it is calculated, the temperature of the stream will not be lowered by the pump to a point where freezing might start to occur.

Though heat pumps are fairly complex devices with their need for precision-engineered compressors, electric motors, valves, tubing and so on, their beauty lies in the fact that they enable us to tap the sun's low temperature energy in a way that no other system makes possible. But the need for compressors, motors, and valves may, in the future, be eliminated by the application of an entirely different set of physical principles to the task of heat pumping.

Heat can be "pumped" from a low temperature to a high temperature by thermo-electrical rather than thermo-mechanical means. In 1834, Peltier observed that if wires made of two dissimilar metals were joined together at both ends to form a loop and if a battery were then inserted in the circuit, one of the junctions would become hot, (indicating that it was giving off heat *to* its surroundings) while the other would become cold (indicating that it was absorbing heat *from* its surroundings). The difference in temperature between the two junctions, he found, was proportional to the voltage of the battery, and was also dependent on the specific metals being used — antimony and bismuth being particularly good performers. Even with bismuth and antimony, however, the actual amount of heat energy transferred from the cold junction to the hot junction was tiny compared to the amount of electrical energy required to "drive" the system.

In recent years, however, the discovery that semi-conductor devices can exhibit much more marked thermo-electric properties than metals has raised hopes that thermo-electric systems using semi-conductors instead of metals may be capable of far greater efficiencies than previously attained — though whether or not it will ever be possible for a thermo-electric device to pump *more* energy from its cold junction to its hot junction than is being supplied electrically is another question. The advantages of such a "solid-state" heat pump would be obvious — no moving parts to go wrong, ease of manufacture, cheapness, silence, and so on.

Semi-conductors also make possible the direct conversion of solar energy into electricity at reasonably high efficiency — ten per cent and more has been achieved in the "solar cells" that have been developed for the US space programme. Such cells depend for their operation on the fact that when light falls on a "semi-conductor junction" (rather like a transistor), it liberates electrons, that form an electric current which may be used directly or stored in batteries. At ten per cent efficiency, a one square metre array of such cells, operating in Britain where the power level of strong sunlight is around 900 watts per square metre, should generate a peak power output of some 90 watts, — but considerably less in diffuse sunlight. If we take the average *annual* amount of sunlight energy received per square metre in Britain to be some 900 kWh, then our one metre square array would generate some 90 kWh a year.

In Britain, silicon solar cells made in the USA by the Solar Power Corporation are now being marketed to the general public by Lucas Aerospace. They measure 18 inches by 14 inches — say one sixth of a metre — and their output in sunlight is quoted as 0.6 amps at 12 volts — or about 7 watts. An array of one square metre area would therefore produce about 40 watts — or, say, 480 watt hours over a 12 hour sunny day. This compares rather unfavourably with the 1,320 watt hours which, as we have seen, is available from a flat plate collector in similar circumstances. Another snag with these panels is their high cost. The Lucas panel just mentioned costs about £200 — which works out at something like £5,000 per kilowatt of "rated capacity", about a hundred times greater than the "cost per rated kilowatt" of a fossil fuel power station. But many electronics laboratories around the world are working on ways to cut the cost and improve the efficiency of solar cells.

One of the most promising developments is a cell using gallium arsenide in combination with gallium aluminium alloys, which has been reported to have achieved 18 per cent efficiency. Indeed the Plessey company in Britain is reported[43] to have developed a particular type of "gallium arsenide — gallium aluminium" cell capable of generating electricity from sunlight at much higher temperatures than hitherto possible. The cell's efficiency, apparently, rises to about 25 per cent when the sunlight is concentrated by means of reflectors of 2,000 times its normal intensity.

Assuming that by concentrating the sunlight some 2,000 times, only one two-thousandth of the number of cells is needed for a given collector area, the savings in cost must be enormous, even allowing for the cost of reflector construction. At 25 per cent efficiency, the peak output in 900 watt per square metre sunshine should be about 225 watts. Some degree of sun tracking, is, however probably necessary, as with focussing solar heat collectors.

Electronics companies are also hopeful that mass production techniques will be able to cut the cost of solar cells by a factor of about 100 — which would cut the cost even of the Lucas-type panel to about £50 per kilowatt, which is highly competitive with the cost per kW of big fossil fuel power plants. The Plessey system should cost even less.

But solar cell manufacture is, at least at the moment, a very highly sophisticated process, capable of being done only by the big international electronics conglomerates. If these multi-national giants do succeed in producing compact, low cost solar power generators and some compact, low cost storage batteries to go with them, they may well be able to "corner" the world solar power market. Which means that the man who runs his lights, fridge and TV set on solar power using say, RCA solar cells is *still* dependent on big, centralised industry for his energy (even if his dependence is not so direct as before because he can do without RCA until his solar panel breaks down and he needs a new one).

Moreover, the mass production of at least some solar cells, if it happens, may not be quite the environmental blessing some would have us believe. As Colin Moorcraft[44] puts it, "Anyone thinking of covering their roof with a nice clean array of cadmium sulphide or gallium arsenide photocells would be well advised to study the effects of heavy metals in the environment."

But solar cells may perhaps turn out to be capable of being manufactured on a much smaller scale than the multi-national monopolists would like. Solar cells are a type of semi-conductor just like the transistors and integrated circuits that have become commonplace in radios, TV sets and domestic appliances. And although transistors and integrated circuits are at the moment almost exclusively mass produced in large factories, many small University laboratories have facilities for their small-scale fabrication: in China, it has been reported that simple transistors are even made in school laboratories.

It is entirely possible that, given the political will to find such a solution, techniques can be developed to enable solar cells to be made in small, worker-controlled factories using a minimum of non-renewable resources, and with no adverse environmental effects. Rare and toxic heavy metals, like cadmium and arsenic, though used in the latest high-efficiency cells, might have to be abandoned in favour of silicon cells which, though less efficient, use as their major raw material one of the most abundant substances on earth.

The potential uses and abuses of solar energy illustrate, however, the important truth that the political and economic system within which a technology is required to operate is just as important in determining its results — adverse or beneficial — as the characteristics of the technology itself. Solar power may be *potentially* a universally-distributed, non-polluting, inexhaustible, free power source, but in a capitalist, (or state-capitalist), exploitative, centralised market economy it can be turned into a commodity like any other — though the task is considerably more difficult than with, say, fossil fuels or mineral resources.

5   Wind Power

If the sun happens to be shining brightly on a cloudless day in a given region of the world, it heats the atmosphere in that region up to a particular temperature, the density of the air falls, and the air pressure (the weight of the 'column' of air above a unit of area) also falls. If, in another region, the weather is cloudy, the sun does not heat the atmosphere to such a high temperature, and the air pressure increases. To equalise the difference in pressure between the two areas, air starts to flow from the area of high pressure to the area of low pressure. That flow is called *wind*.

The theoretical amount of power that can be generated by wind of a given speed blowing on a given surface, such as a sail, can be calculated fairly readily.[45]

In summary, the power available from a windmill can be shown to be proportional to the square of the windmill's diameter, and to the cube of the wind speed. The actual power output, say P_{act}, is given by the formula:

$$P_{act} = \frac{35}{100} \cdot 510 \cdot 10^{-6} \cdot D^2 \cdot V^3, \text{ (in kilowatts.)}$$

Most windmills have a "rated" windspeed for which they are designed and at which they work at maximum efficiency — usually between 20 and 35 miles per hour (say, between 10 and 16 metres per second). So if we have a windmill of, say 12ft diameter (roughly four metres), in a wind speed of, say 10 metres per second, the amount of power we can expect to generate is some 2.85 Kilowatts.

Now of course, wind does not blow constantly for anything like all the 8,760 hours in a year — if it did, one could expect to be able to generate some 25,000 Kwh of energy a year from such a mill. In practice, the actual amount obtained by a given windmill in a given location depends to a tremendous extent on the exact characteristics of the location.

Statistics like the *average* windspeed per year in a particular site are of little help in calculating what the total output might be, however, because an average figure may conceal vast differences in the actual conditions. The only way to calculate the amount of power to be expected from a given windmill on a given site with any degree of accuracy is to take wind speed measurements over a full year and draw a *velocity-duration curve* — a graph which shows the number of hours a year for which the wind blows at various velocities. By then cubing the values for velocity (because the power available is proportional to the velocity cubed) and allowing for the fact that the windmill does not operate at all below a certain wind speed (say below 6 mph or 3 metres per second) and that it must be taken out of action if the winds are too strong (about 60 mph or 27 metres per second), it is possible, as Golding[46] shows, to obtain a curve that shows the variation of power generated with the number of hours in the year. By taking the area of this curve, one obtains the total amount of energy obtainable per year from the windmill in the given conditions.

There are, however, some rules of thumb about windmill power which are quoted from time to time, and which may or may not be misleading. One yardstick is that a windmill can give about 1200 kilowatt hours a year for every kilowatt of output at its rated windspeed. By that yardstick, our 4 metre windmill of 2.85 kW rated output ought to be able to generate some 3,500 kilowatt hours a year — not far short of the annual domestic electricity consumption for all purposes in the average UK household. An "aerogenerator" mill of similar characteristics to this is available in the UK, in the shape of the Dunlite BR Mill, made in the USA and available from Conservation Tools and Technology Ltd[47] (formerly Low Impact Technology Ltd.) which has a 3.5 metre diametre, three-bladed propellor, and has a rated output of some 2,000 watts. It should, therefore, be able to provide some 2,400 kWh a year, by the above criteria.

One drawback, however, is the price. At some £1025 excluding tower, storage batteries, and control circuitry, power from such a mill is pretty expensive, even with CEGB electricity prices soaring. By calculating the annual cost of a typical aerogenerator over its estimated lifetime, Gerry Smith of Cambridge University has worked out[48] that electricity from a windmill installation like the one described is about *twice* as expensive as the same amount of main-supplied electricity, even without storage facilities. If storage batteries are required, the cost per kilowatt hour can rise to *six times* as much. Clearly, wind-generated electricity should only be used for those purposes for which no other energy source will suffice — such as lighting, running TV sets and radios, and powering some domestic appliances. In a more optimistic vein, however, Smith also calculates that if a small windmill is used solely to supply house *lighting* requirements, and if "caravan-type" fluorescent lights (with built in DC-to-AC transistorised invertors) each with an associated storage battery, are employed, then the cost of lighting per "effective" kilowatt hour is about the *same* as the normal cost of home lighting in which ordinary incandescent bulbs and mains electricity are used. This is because fluorescent lights are five to seven times as efficient at light production as incandescent lights. (But of course one could equally well run one's fluorescent lights off the mains).

There have, however, been some ominous suggestions that fluorescent lights (and colour TV sets incidentally) emit X-rays and similar radiations in doses that could be harmful to human beings.[49] Another negative point about fluorescent lights is that the "daylight white" types of tube do *not* emit the wavelengths of light that normally help build up vitamin D in human tissues — though this is not the case with "warm white" types.

Lighting companies and most straight scientists sneer at the idea that such "low level" radiation can be harmful. But it could well be that such "experts" are as wrong about fluorescent lighting as they have been about "low level" nuclear radiation. It is, however, possible to shield fluorescents by wrapping a few inches of lead foil around the cathodes, and encasing the entire fitting in a metal screen (with holes to let the light out).

*Windmill types

There are a number of different windmill classifications, but perhaps the simplest is that which divides windmills up into

*(All prices mentioned in this section were operative in 1974)

61

two types: those which have a horizontal axis, and those which have a vertical axis.

Horizontal axis windmills are by far the most common types. Like the old windmills that still grace the landscape in Holland and a few parts of Britain, they consist of sails which move in a direction perpendicular to the wind which falls on them, and which are pivoted about a horizontal axis. Because the wind direction varies from hour to hour and day to day, such windmills need a steering mechanism to make sure that the axis around which their sails rotates always faces into the wind. These large, graceful machines are, by today's standards, not very efficient and very costly to build. When people talk of windmills nowadays, they usually think of the relatively small "aerogenerators" which farmers in remote regions of the USA and the Australian outback still use for generating electric power, when no suitable mains supply is available at reasonable cost.

One typical aerogenerator is the Dunlite windmill[50] (referred to above), which has become popular throughout the world. Two versions are available — one with 1200 watt rated output, costing £950, the other with 2,000 watt rated output, at £1025. Similar mills, made by the Electro company of Switzerland, are on sale through CTT Ltd at prices ranging from £495 for a 600 watt-rating model of 2.5 metre diameter, to £1125 for a 4,000 watt machine of 4.4 metres diameter. CTT also sells a small, 200 watt "Wincharger" for around £200. (See illustration on page 63)

Most of these machines give direct current electricity, as opposed to the AC current provided by the normal mains, and most of them deliver voltages that are lower than the 220 volts or so that is standard in this country. Many deliver 12 volts DC, some 24 volts DC, others the American 110 volts.

Direct current, of course, is perfectly acceptable for running many household devices. Light bulbs (including fluorescents with a little modification) can use it, as can transistor radios and many types of TV set (at least those without a transformer). But devices with motors or transformers installed, unless especially designed to work on DC, cannot be used directly. If it is desired to run such devices from a DC

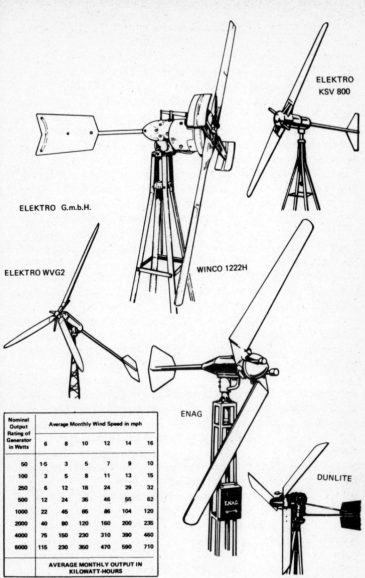

Nominal Output Rating of Generator in Watts	Average Monthly Wind Speed in mph					
	6	8	10	12	14	16
50	1·5	3	5	7	9	10
100	3	5	8	11	13	15
250	6	12	18	24	29	32
500	12	24	35	46	55	62
1000	22	45	65	86	104	120
2000	40	80	120	160	200	235
4000	75	150	230	310	390	460
6000	115	230	350	470	590	710
AVERAGE MONTHLY OUTPUT IN KILOWATT-HOURS						

Some of the wind generators available from Conservation Tools and Technology Ltd.

source, the motor must be either re-wound or replaced with a DC equivalent, which is not always possible.

Alternatively, the DC output of the aerogenerator can be turned into AC by means of a device called an "Invertor". Invertors are of two types — static and dynamic. The "dynamic" type of invertor usually consists of a motor running on DC current, coupled directly to a dynamo which generates AC current. Such "rotary" invertor can be picked up pretty cheaply (say, under £10 for a 250 watt output device — see ads in *Exchange and Mart*) as ex-government surplus, but do not have a very high efficiency. They are also somewhat unreliable. Only some 40 per cent of the DC energy from the windmill will appear as 250 volt AC energy at the output of a rotary invertor.

Another variety of invertor, whose classification lies somewhere in between the dynamic and static categories, is the "vibrator" type. In these devices, the incoming DC current is converted to a crude form of AC by means of a "vibrator" (rather similar to a doorbell or buzzer) and then stepped-up in voltage by a transformer. An efficiency of about 65 per cent is typical.

Static-type inverters, however, have, as their name implies, no moving parts at all, and can achieve an efficiency of 80 to 90 per cent. The direct current input to such devices is used to drive a high-power transistor oscillator circuit (usually working at 50 cycle mains frequency, except in the case of the small static invertors used to power "caravan-type" fluorescent lights, which work at a higher frequency and often give off a high-pitched whine which can be annoying). The alternating current from the oscillator can then be stepped up, using a transformer, to 250 volts. But to derive high powers from static invertors needs a lot of high-power transistors, and these are expensive. A 150 watt invertor costs a fairly-reasonable £25, but a 2,000 watt invertor has a cool £750-plus price tag.

Power storage is another expensive aspect of the wind-electricity business. A typical 12 volt car battery — of the *lead-acid* type — might have 50 ampere-hours of storage capacity. This means that, when full, it can deliver a current of 50 amperes at 12 volts for an hour before being fully discharged. Since energy in watt-hours equals voltage multiplied by current multiplied by time, such a battery can store 12 . 50 . 1 watt hours (0.6 kilowatt hours) of energy.

The number of charge-discharge cycles which a battery can handle before its performance deteriorates is obviously another important consideration. An average car battery has a lifetime of only 750 to 1200 cycles. At, say, one charge-discharge cycle a day the lifetime is therefore 2 to 3 years. (Nickel-Cadium batteries can give up to 2,000 cycles lifetime, but cost about 3 times as much). CTT recommends the use of heavy duty seagoing lead-acid batteries, which cost about £40 for a 125 amp-hour 12 volt model (having 1.5 kilowatt hours of storage capacity). For the small 200 watt output Wincharger, CTT recommends 250 A.h. of storage (3kWh), which would cost £80. For a 1-2 Kw output Dunlite, CTT recommend 200 to 400 A.h. of storage, costing say £80 to £120. But the costs don't stop there.

A tower to mount your windmill on, even if you use an adapted GPO telephone pole, will set you back £50 to £75. And cable, at £1 a yard for not more than 75 to 125 yds (if you use a longer cable, the resistive losses can become unacceptably high), is, say, another £100.

To give an idea of the *total* cost of a commercially-installed wind power installation, therefore, let's take two examples (all prices exclude VAT and carriage, by the way). At the low-cost end of the scale, say one were to decide to instal a low-power 200 watt Wincharger to power 10 low-power fluorescent light tubes, as CTT suggests. The Wincharger costs £175. Two seagoing batteries (250 A.h capacity) cost £80. Ten low voltage fluorescents, with built-in frequency transistorised invertors, cost £5 each = £50. The total so far is £305 — though CTT will do a "package deal" (including only one heavy duty battery) for only £245. Add on to that, say only £50 for a tower, £75 for cable, and say only another £25 for miscellaneous switches, plugs, and so on. The total comes (even taking advantage of the CTT package deal) to £395 — say £400 (excluding VAT and carriage. And if the Wincharger, with its 200 w output at rated speed, delivers the rule-of-thumb 1200 Kwh per kilowatt, then it will supply say 240 kWh of electricity per year. The cost of such electricity at present prices is about 1 to 2 p per Kwh, depending on whether one takes into account the "standing charges". At 2p per unit, 300 kWh of power would cost about £5 per year if supplied from the mains. So the unit would take something like 80 years to pay for itself — assuming, in a rough-and-ready approximation, that the interest which would otherwise accrue on the £375 capital, plus the cost of maintenance and replacing batteries every three years or so, is cancelled out effectively by the escalating cost of mains electricity. Let's take another, more powerful example. A 2 kW Dunlite aerogenerator costs £1025. Assuming the following additional costs — £75 for a tower; £100 for cable; £50 for switching; £125 for batteries; and £25 for a 300 watt static invertor — the total price works out at £1400.

Assuming also that a 2 Kw-rated mill will deliver 2,400 kWh of power a year, the cost of that power from the Electricity Board would be £48. So, again assuming that the effects of interest charges, battery replacements and maintenance costs are cancelled out by electricity price inflation, the system would take nearly 30 years to pay for itself. It is clear that such costs can only be afforded by those who value very highly their independence from the CEGB and who are unconcerned about their dependence on the battery and windmill manufacturers. Equally obviously, the only way to cut the cost of wind power is to make it yourself — at least until such time as mass production lowers the cost of windmills by at least an order of magnitude (10 times). And if the latter happens, one has to consider at what cost, in social terms, has that mass production been achieved: are the windmills being churned out cheaply by slave labourers in Taiwan, or being put together by craftsmen in a workers co-operative in Chipping Norton? It is, of course, possible to *build oneself* an equivalent of, say, the Dunlite or Wincharger mills at a fraction of the cost. The "propellor" you can carve from wood. The tower can be a wooden telegraph pole or of metal construction. There are a number of excellent do-it-yourself designs to choose from[51]. Many of these designs employ old car and bicycle bits, to cut the cost of the precision-made parts.

One low-powered but particularly easy design uses a standard bicycle wheel with a built-in Sturmey Archer

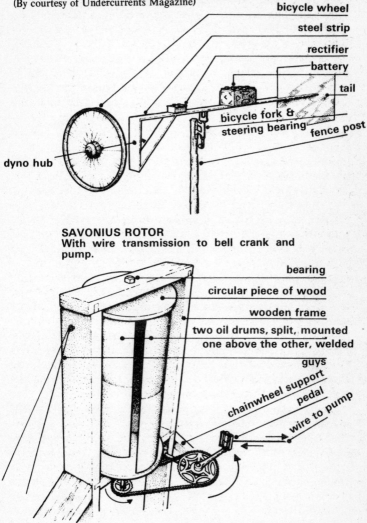

Low Powered wind generator made from bicycle parts. The windmill blades (not shown) are made from eight strips of aluminium attached between rim and hub and twisted to an aerodynamically-favourable shape.
(By courtesy of Undercurrents Magazine)

bicycle wheel

steel strip

rectifier

battery

tail

bicycle fork & steering bearing

fence post

dyno hub

SAVONIUS ROTOR
With wire transmission to bell crank and pump.

bearing

circular piece of wood

wooden frame

two oil drums, split, mounted one above the other, welded

guys

chainwheel support

pedal

wire to pump

Dyno-Hub generator. The wheel is fixed to a pole with a vane at the other end, and the whole assembly is pivoted on a pair of cycle front forks! (See top illustration on page 67).

But because of the need to pivot the propellor in horizontal-axis windmills, and because of the need for precisely-shaped aerofoil surfaces (though sail-type windmill blades have surprisingly high efficiency) and also because of the consequent need for "slip-rings" to carry the current away from the dynamo to the ground, the easiest type of windmill to build is a vertical-axis machine, which, at a stroke, eliminates the need for such measures. And the easiest vertical axis machine of all to build is the Savonius Rotor. The most famous Savonius design is that developed by the Brace Research Institute in Canada, which uses a 45 gallon oil drum, sawn in half to form the two cup-sections that are characteristic of the Savonius system. (One variation of this is show at foot of p67).

In the Street Farmers' version of the Brace design, the two halves are placed off-set from each other (see illustration on page 69) and joined at the top by being bolted to the chain wheel of a bicycle. The bottoms of the two halves are joined by being bolted to a large, 4ft diameter, circular piece of plywood, which acts both as a flywheel to even out any irregularities in rotation, and as one element in an ingenious gearing system. The Savonius Rotor rotates relatively slowly compared to aerodynamic horizontal-axis machines (this does not necessarily mean that the *power* is any lower, however). If the rotor is to be used for electricity generation — as opposed to, say, water pumping — using a car dynamo or alternator, its rotation has to be geared to suit the fairly high rotational speeds at which these generators operate. A car dynamo, for instance, begins to charge at about 900 revolutions a minute, and an alternator charges at about 600 revolutions a minute. The Street Farmers' Savonius uses the wooden flywheel as the large "gear" of a friction-type gearing-up mechanism. The alternator hub is fitted with a rubber wheel and the alternator is mounted on a swinging arm which allows the rubber wheel to impinge on the surface of the flywheel — which in turn is coated with heavy duty sponge rubber. Power at 12 volts is then taken away for use as with a normal windmill. The Savonius

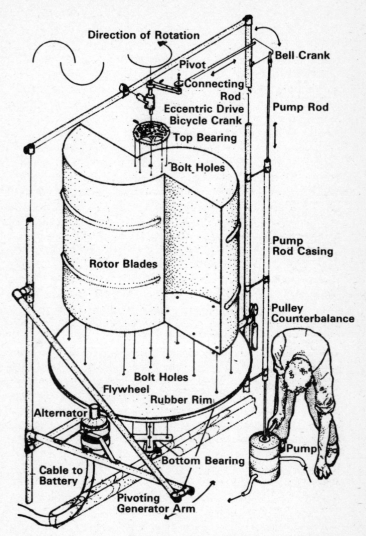

Direction of Rotation

Pivot

Connecting Rod

Eccentric Drive
Bicycle Crank

Top Bearing

Bolt Holes

Bell Crank

Pump Rod

Pump
Rod Casing

Pulley
Counterbalance

Rotor Blades

Bolt Holes
Flywheel

Rubber Rim

Alternator

Pump

Cable to
Battery

Bottom Bearing

Pivoting
Generator Arm

(By courtesy of Bruce Haggart, Street Farmers)

69

can also be used very easily to pump water, as shown in the diagram. Indeed, the fact that most people nowadays think of windmills in terms of electricity generation should not be allowed to obscure the useful *mechanical* work capable of being done by wind power. Wind power can even be used to generate heat directly, by using it to generate deliberate frictional heating in a brake. Joule demonstrated the mechanical equivalent of heat in this very manner — though he did not use a windmill to provide the mechanical input. Windmills can also be used to drive heat pumps directly, a technique with very considerable potential, as we shall see.

It is instructive to compare the cost of a home-built Savonius Rotor wind generator, of the Street-Farmer/Brace Institute type, with the cost of buying ready-built units. Let's allow, say, £25 for the alternator (new: less if second-hand); £5 for the oil drum; £10 for the wooden framework; £10 for the bicycle and other bearings; and say another £10 for the pump and associated mechanism, if desired.

A tower might not be necessary if the unit were mounted on the side of a house or between two houses (see illustration on page 71) but let's assume the D.I.Y. cost is, say, £50 (for perhaps 2 telegraph poles, cables and fittings etc.). As for batteries — let's say one were to need 250 A.h of storage at 12v (3 kWh) using reconditioned car batteries of 50 A.h capacity (second-hand, *un*-reconditioned car batteries are worse than useless!) at £7 each, the cost would be £35. Let's be pessimistic and assume that 75 ft of cable would cost the full £75 as in the Wincharger example above. And let's add £25 for a 300-watt invertor, plus £25 for switchgear, ammeter, etc. The bill comes to £270 (though we can cut out the £50 for the tower if the unit is gable-mounted, plus say £50 of the £75 cable costs because of the short run from gable to house. The total then comes to £170).

What power would such a unit be capable of producing? The area swept by the "blades" is 4′ high by 4′ wide — say 2 square metres (ignoring, for simplicity, the "Magnus Effect" — an aerodynamic phenomenon which effectively makes the Savonius Rotor *more* powerful than its area would suggest).

Artist's impression of two Savonius Rotors mounted vertically above one another between two houses. (By courtesy of Bruce Haggart, Street Farmers)

Using the windmill formula

$$P = \frac{35}{100} \cdot \frac{(.65 \cdot A \cdot V^3)}{1000}$$

and assuming a "rated" windspeed of 10 metres/sec., the power output becomes

$$P = \frac{35}{100}(.65 \cdot 2)$$
$$= 430 \text{ watts.}$$

Assuming 1200 kilowatt-hours per year of energy, per rated kilowatt, this means that the unit should deliver some $\frac{1200 \cdot 430}{1000}$ kWh, or about 516 kilowatt hours a year.

To supply this power from the mains would cost some £10.

At the £270 capital cost assumed for the full system, it would take about 27 years to pay for itself. (At £170, it would take 17 years.) Assuming good design and strong construction, there is no reason whatsoever why the system should not last 20 years — there is so little to go wrong.

71

"Darrieus" type windmill, developed by Raj Rangi and Peter South. (By courtesy of NASH)

A much more sophisticated vertical-axis machine than the Savonius was developed a couple of years ago in Canada by Raj Rangi and Peter South, two researchers at the National Research Council of Canada's Aeronautical Research Establishment.[46] Rangi and South's windmill (becoming known as a "Darrieus" wind machine, because it is based on principles developed by a French engineer of that name in the 1920s) looks like a helicopter rotor which has had its blades bent upwards in a circle, so that they meet again at the tips as well as being joined to the rotor hub. (See illustration on page 72).

As Ranji and South describe it:

"The Rotor consists of 2 or 3 convex metal blades of aerofoil cross section attached to a vertical shaft, supported on ball bearings at the top and bottom of the shaft, and held with guy wires at the top".

The beauty of the Darrieus machine is its ability, because of the aerofoil shape of its blades, to generate power in high wind speeds — something a Savonius Rotor is not very efficient at. It is also very simple to make — almost as simple as the Savonius. One disadvantage, however, is that it has a very high "starting torque", which means that it needs to be "spun up" before it will start to behave efficiently in a stiff breeze.

But the best of both worlds can be had if one combines the high-speed Darrieus with the slow-speed Savonius, by placing a Savonius at the centre of the Darrieus 'loop'. (See illustration on page 74). The Savonius would work efficiently in light winds, and would provide the starting torque for the Darrieus to take over in high winds. Given the dimensions of the original prototype (15ft, or say 5 metres, diameter) one could expect in 10m/sec. winds a power output of something like

$$P = \frac{35}{100} \frac{(510 \cdot 25 \cdot 1000)}{106} = 4.16 \text{ kilowatts}$$

And assuming the unit's ability to work in high *and* low windspeeds, it should be able to deliver at least our "guesstimated" 1200 kWh per kw per year — that is, a total of over 5,000 Kwh.

Let us assume the DIY construction costs of such a device, (including tower, storage batteries, etc. etc. . . . .) to be, say, twice as much as those of the Savonius Rotor just mentioned — say £500. The cost of supplying 5,000 kWh a year of electricity is £100 — so the unit could pay for itself in about 5 years, *and* deliver the large amounts of electricity currently being demanded by average householders. (Alternatively, and preferably, a smaller unit costing less could deliver more sensible amounts of energy).

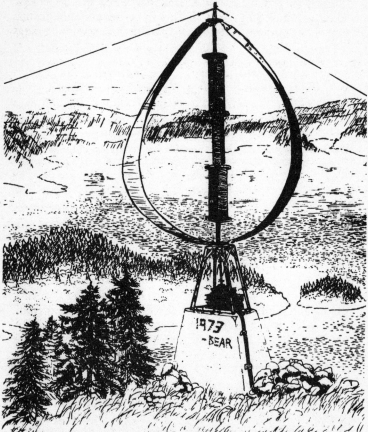

A "Darrieus" vertical axis windmill, designed for high speed winds, can be "started" by a low speed Savonius Rotor placed at its centre. (By courtesy of Kenya Lee: The Natural Energy Workbook)

6 Water Power

Whether or not you can take advantage of the power of running water depends, of course, on whether or not you have access to a stream or river. This highly important qualification sets water power apart from the other renewable energy resources dealt with so far. Direct solar power, plant power and wind power can all be tapped in virtually any location — though with widely varying degrees of effectiveness.

The total amount of hydro-electric power generated in Britain in 1970 was about 5×10^9 kilowatt hours — about one-fortieth of the total amount of electricity generated, some 200×10^9 kilowatt hours. And 1970 was a good year. In 1971 and 1972, presumably because of lower annual rainfall, less than 4×10^9 kilowatt hours was produced by hydro-electric means.

But according to Andrew MacKillop,[53] official estimates of the amount of potential hydro-power available in Britain are much too pessimistic, because they are based on the arbitrary assumption that hydro-power plants under 250 kW capacity, served by water catchment areas of less than 500 square kilometres, are "uneconomic".

As MacKillop points out, even small streams are capable of generating surprisingly large amounts of power. He quotes the example of a typical stream with less than $50m^2$ catchment area, and a minimum flow rate of only 0.2 metres per second. Even this flow rate, if the water is allowed to fall through a modest height of 2 metres, will give a minimum of 1.6 kilowatts of electric power, even allowing for more than 50 per cent losses due to turbine and generator inefficiency, friction, and so on. Over a year of 8,760 hours, such a stream would give over 14,000 kilowatt hours of energy. MacKillop concludes that the flow of small and medium-sized rivers could, if tapped, raise Britain's hydro-electric power output to some 40×10^9 kWh a year. And whether or not one agrees with that particular estimate, it does seem obvious that a great deal of Britain's hydro-electric power is going to waste.

A river, as a moving body of water, possesses energy in two characteristic forms: it has "kinetic" energy due to its motion — the amount of kinetic energy flowing per second being equal to half the mass of water multiplied per second by the square of the water's velocity; and it has "potential" energy because the mass of water, even if it were stationary, would be able to do work if it were made to fall over a distance. The potential energy that can be released by a falling mass of water per second, is, as one might intuitively expect, equal to the weight of water per second multiplied by the height (or "head") through which it falls.

The various water power devices which have been developed over the centuries can therefore be divided into two main categories: those which utilise the water's *potential* energy; and those which utilise its *kinetic* energy. There is, of course, a third category: those which utilise *both* forms of energy.

The so-called "undershot" water wheel (see top illustration on page 77) — like a paddle steamer wheel used in reverse — uses mainly the kinetic energy of a stream.

To utilise the potential energy of a stream, one must naturally allow the water to fall through a distance. This can be accomplished either by building a dam — which allows the water to build up to a height greater than normal before it falls back to its original level; or by channelling some or all of the water in the stream through a pipe or aqueduct, often called a *penstock,* down to a location at a lower altitude than the point at which the water enters the pipe; or by a combination of the two methods. Obviously the greater the height of the dam the greater the potential energy that can be gained when the water "falls over" it. Equally, the greater the vertical distance between the point at which water enters a penstock and the point of exit, the greater the potential energy.

One type of water wheel that uses *only* the potential energy of falling water is the "breast mill" (See centre illustration on page 77), so called because it incorporates a curved "breast" close to the wheel, to prevent water spilling out from the buckets before it has done the maximum amount of work. There is some overlapping between the "undershot" and "breast"

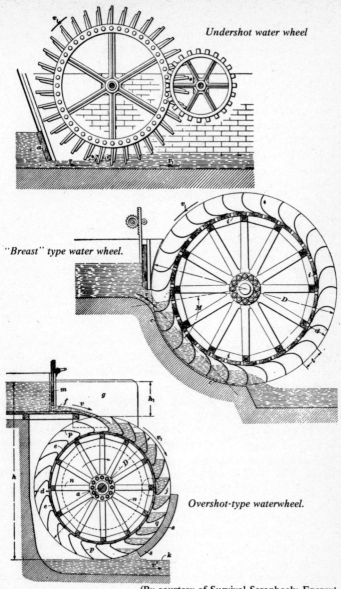

Undershot water wheel

"Breast" type water wheel.

Overshot-type waterwheel.

(By courtesy of Survival Scrapbook: Energy)

categories, since some undershot wheels utilise the potential energy of a head of water, as well as its kinetic energy.

Another potential energy-actuated wheel is the "overshot" type. (See bottom illustration on page 77).

Water *wheels* are only one way in which water power can be harnessed. A more modern — though not always more efficient — method is to use a *turbine*. Turbines, like water wheels, are of two kinds: those which use the kinetic energy of the water, which are called *impulse turbines;* and those which use the potential energy of the water, which are called *reaction* turbines. (To complicate matters, however, there are some designs which use both the kinetic and the potential energy components of the water flow.)

One type of turbine that employs only the kinetic energy of water is called a "Pelton wheel" (or often just an *impulse* wheel) (See illustration on page 79). The Pelton wheel is sometimes called a "developed undershot" wheel, because of its similarity to the ordinary undershot wheel. It incorporates a nozzle at the exit of the penstock which increases the velocity of water flow. This means that the wheel turns faster than other wheels — which is useful for some purposes, such as power generation.

In the *reaction* turbine, water is guided by stationary vanes on to angled moving vanes which rotate as they *react* to the weight of water falling on them. There are many different types of reaction turbine —'Fourneyron,' 'propellor-type', and so on. Some turbine wheels, as mentioned earlier, utilise both kinetic and potential components of the water energy: examples are the Mitchell (or Banki) turbine (See illustration on page 80), and the Francis (or American) turbine. The amount of power that can be generated by a given water wheel or water turbine depends both on the efficiency of the device itself and on the amount of kinetic and potential energy available. George Woolston[54] has summarised the pros and cons of some of the various types as follows: "The overshot wheel uses heads of 10 to 30 feet and flow rates of 1 to 30 feet per second, and gives an efficiency of between 60 and 80 per cent. The undershot wheel uses heads of 1.5 to 10 feet and flow rates of 10 to 100 feet per second, and gives an efficiency between 60 and 75 per cent.

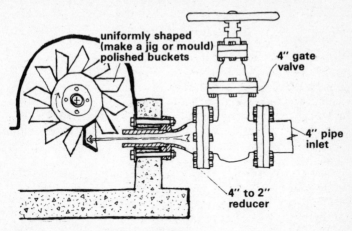

**uniformly shaped
(make a jig or mould)
polished buckets**

**4" gate
valve**

**4" pipe
inlet**

**4" to 2"
reducer**

Home made Pelton wheel water turbine.

(By courtesy of Survival Scrapbook: Energy)

The developed undershot or impulse-wheel (a turbine) has an efficiency of 80 to 90 per cent or more, and is used for high heads and low flow rates."

In the case of an undershot wheel spinning directly above a stream and utilising only its kinetic energy, without the use of a dam or penstock to increase the "head" of potential energy, the power available must be calculated by measuring the speed V of the river flow, and the volume in cubic feet of water per second passing through the vanes of the wheel. Multiplying the number of cubic feet per second by the density of water (62.5lb/cu ft) gives the mass M flowing per second. And multiplying half the mass by the square of the stream velocity (i.e. $\frac{1}{2}MV^2$) gives the energy available per second.

Except in the case of such an undershot wheel, though, the kinetic energy component in the water flow is usually ignored because the dam slows the water flow down almost to zero just before it reaches the edge of the dam, at which point its kinetic energy is negligible. (But in the case of a turbine fed from a pipe or penstock which draws its water directly from the river

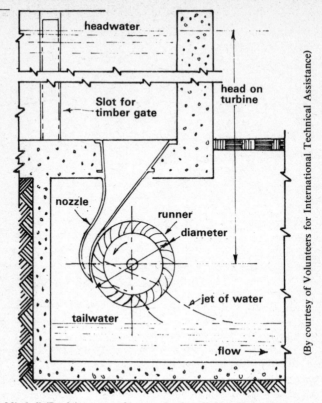

The Mitchell (Banki) water turbine, a simple, efficient design that can be assembled in a modest workshop of the type used to repair farm machinery.

and not from a dam, the kinetic energy of the moving water before it enters the pipe should be added to the potential energy.)

Ignoring the kinetic component, however, the power available from a given flow of water (in, say, cubic feet per second) falling through a given "head" (in feet) can be calculated by multiplying the volume per second by 62.5 (the weight in lbs of one cubic foot of water), and then multiplying the result by the head. The answer, in foot-pounds per second, is the power available. To convert the power to a more familiar measure, horsepower, just divide by 550, which is the number of foot pounds per second in one horsepower. And to convert to kilowatts, multiply the above result by 0.748, since one horsepower is 0.748 kilowatts. Thus Power (in kilowatts) equals

$$\frac{62.4 \cdot V \cdot H \cdot 0.748}{550}$$

On applying this formula, it can be seen that the amounts of power available in even the most modest-seeming streams are quite considerable. Take, for example, a stream whose flow rate, averaged throughout the year, is say 1 cu ft per second, with a dam which provides a modest 6ft of fall. The power available is:

$$P = \frac{1 \cdot 62.4 \cdot 6}{550} \cdot 0.748 \quad (kW)$$

i.e. about ½ kW

Even allowing for 50 per cent losses in the turbine, and in generation and transmission, that still leaves about ¼ kW — which may not seem a lot. But let's not forget that even ¼ kW (given the above assumption that the *average* flow throughout the year is 1 cu ft per second) is equivalent to ¼ x 8,760 kWh, per year or about half the annual average electricity consumption per household, and more than enough to run domestic lights and most appliances.

Another point about seemingly small powers like ¼ kW is that a dam can provide water *storage* facilities — which means that the water can be allowed to build up to a level which, when released, will provide a much faster flow rate than normal, so allowing much more than ¼ kW to be used for limited periods. In almost all households, demand is not constant and such a storage system, if accurately matched to the pattern of demand, would probably be quite capable of handling peak demands, provided that the *total* annual electricity demand did not exceed the total supply. As regards the *cost* of specific water power installations, these depend, to an even greater extent than is the case with other renewable energy sources, on the *exact* circumstances — the size of water wheel or turbine, the size of dam, the length of penstock, the need for buildings to house generators and switchgear, plus labour costs if the project is not being done on a Do-it-Yourself basis.

Purely as a 'guesstimate', let's assume a small home-made

81

A ship mill can enable water power to be harnessed without the need for constructing a dam. (By courtesy of George Woolston)

dam across the very small, 1 cu ft per second, stream considered earlier, driving a home-built overshot wheel situated just below the dam, which drives a DC generator coupled by short cable (say 50 yards) to a house where there is, say, a 300W invertor to drive non-DC appliances. Let's say the dam might be able to be made for, say £100 in materials — wood, sand-bags, and so on. Let's say the wheel could be made from two old cart-wheels and other bits and pieces — say at a cost of £50 in all, including bearings and gearing. (Another unorthodox starting point for the construction of a water wheel might be a discarded wooden cable drum — though I don't know of anyone who has tried it). Add £50 for the generator, £50 for cable, and £30 for the invertor, and perhaps another £100 for the storage batteries, switching, regulators and so on. The total comes to about £380.

The Power generated per year, as calculated above, is 2,200 kWh, which would cost, at present prices, say 2p per kWh or, say, £45 per year. Assuming electricity cost inflation balances the interest lost on £380 capital, the system should pay for itself in about 8 years. The construction of a dam, normally regarded as a prerequisite for tapping the power of flowing water, is not strictly necessary, however. So-called *Ship Mills*, (see illustration on page 82), rather like permanently-moored paddle steamers, enable the energy of a river, or of a tidal flow, to be harnessed with much less disturbance and inconvenience than is necessitated by dam construction. The ship mill (or "floating" mill) consists of one or more undershot water wheels mounted on a ship, which is moored in the middle of a river or tidal estuary. The river turns the wheels, generating power which is normally utilised on board the ship — say for corn milling or in running a paper mill or saw mill.

Such mills are still in use in Romania[55]. More modern ship mills could be used for electric power generation — the power being conveniently brought ashore via the tethering cables. Other ingeniously simple possibilities that could be explored include the water-driven ferry which, as George Woolston explains, "is a ferry which is tethered by a long rope from a point upstream to a capstan on board. The current rotates the paddle wheels which in turn drive the capstan to pull the ferry upstream. Downstream travel is by drifting with the current." As Mansur Hoda[56] puts it: "any clever fool can make things complicated; it takes a touch of genius to make them simple again."

7 Integration

The second law of thermodynamics implies that there exists a hierarchy of energy sources, some more *useful* than others. Gravitational (or "potential") energy, for instance, is more "useful" than mechanical energy, because gravitational energy (in the form of falling water) can be turned into mechanical energy (say, the rotation of a turbine) with almost one hundred per cent efficiency, but the conversion of mechanical energy to gravitational energy is a less efficient process.

Mechanical energy, in turn, is more useful than electrical energy, because mechanical energy can be changed into electrical energy (in a dynamo) with almost one hundred per cent efficiency, but the efficiency of conversion of electrical to mechanical energy (in an electric motor) is somewhat lower. Lower still on the scale is heat energy, which can be generated very efficiently from mechanical energy (in a friction brake, for example), or electrical energy (an immersion heater), but which can only be converted back into these forms (in a steam engine, for instance) at a relatively low efficiency. Heat itself varies in usefulness, high temperature heat being, as one might expect, more "useful" than low temperature heat.

What these thermodynamic facts of life mean is that although all energy forms are "equal", in the sense that they are all theoretically capable of performing an equal amount of work for a given quantity of energy, some are more equal than others, in that their qualitative effectiveness varies.

Any attempt to match the characteristics of sun, wind and water power to the demands, say, of a domestic household for energy to run space heating, lighting, cooking, motor-driven appliances, and so on, should therefore take into account the suitability of the various energy sources for the tasks which they are to perform. To take a few examples: the most immediate and direct result of wind, for instance, or of the flow of a river, is mechanical energy, so wind and water power are most suited to providing direct mechanical work. This is how wind and water mills were used in their original forms — to grind corn, to saw wood, and to pump water. Their more recent use in electricity generation is, in theory, slightly less efficient than their use for mechanical purposes, but in practice this factor is often outweighed by the greater flexibility of electrical energy.

Solar power, on the other hand, is most directly suited to providing heat — and although solar cells are capable of turning the sun's radiation into electricity, their efficiency is fairly low. The solar energy stored by plants as the result of photosynthesis is obviously best suited for use in providing *food* energy. But if we want to capture the energy for non-food purposes, the next best thing is to digest the plants anaerobically to produce methane fuel. Methane plays a key role because it provides transportable, stored energy in a convenient form. The only disadvantage of methane is that, in gaseous form, it takes up quite a lot of space for a given amount of energy, and that in order to turn it into a more compact liquid form at normal temperatures it must be put under very high pressure, which requires a considerable amount of compressive energy, and necessitates the use of heavy metal "bottles".

The other forms of energy can, of course, also be stored. As we have seen, heat can be stored in an insulated tank, or in a thick concrete wall, or by using the latent heat properties of chemicals like Glauber's salt. Mechanical energy can be stored by means of a spinning flywheel.

Electrical energy can, of course, be stored in chemical form, in batteries.

These and other forms of energy storage enable the variability of most of the renewable energy sources to be offset to some extent, depending on the amount of storage.

But another way of getting a more or less constant supply of renewable energy is to take advantage of the *complementarity* of at least some renewable sources. Winds, for instance, are usually strong in winter when there is little sun, and fairly gentle in summer when sunshine is plentiful. To employ energy sources in a complementary fashion, however, means using

A model of one of the autonomous House designs developed at Cambridge University, under a project financed by the Science Research Council.

(By courtesy of John Donat Photography)

more than one source for each potential application, which conflicts to some extent with the principle of using only the most appropriate source for each application. It could mean, perhaps, using high-grade, wind-generated electricity for low-grade heating purposes on particularly cold, but windy, days in winter.

It is, for all these reasons, almost impossible to lay down hard and fast rules specifying which particular renewable energy source should be used in which specific task. The tailoring of renewable energies to suit human requirements is likely to remain more of an art than a science. The creation of a harmonious solution to a set of energy requirements will always depend as much on the intuitive skill of the craftsman-engineer as it will on the scientific principles involved.

The autonomous housing project (See illustration on page 87) at Cambridge University incorporates quite a lot of "synergistic" thinking of this kind. For instance when Gerry Smith, one of the Cambridge team, was faced with his own research conclusion that electric power from an aerogenerator costs between two and six times as much as power from the national grid, he came up with a neat solution to the problem: Smith's suggestion was that mechanical power, taken directly from a small windmill, could be used to drive the compressor in a heat pump. This move effectively multiplies the energy output of the windmill by the coefficient of performance of the heat pump — a factor of about three — which effectively cancels out the cost disparity between wind-generated electricity and its mains equivalent, at least as far as heating purposes are concerned.

What's more, Smith suggests that the *source* of heat for the wind-powered heat pump, instead of being the conventional medium of air, soil or water, should be the solar collector which it is planned to incorporate in the autonomous house. In summer, the solar collector will provide energy directly for water heating, but in winter, as we have seen, there is not enough sunshine in Britain to give a high enough water temperature for washing purposes. This does not mean, however, that *no* energy is available — it's just that the temperature isn't high enough. But the heat pump is capable of transforming unusable low-temperature heat into useful high temperature heat. (See illustration on page 53).

Smith's proposal (though it has still to be tested in practice) illustrates how the inherent characteristics of renewable energy sources can complement one another, if the systems chosen to harness them are designed creatively. Colin Moorcraft's and the New Alchemists' proposals to integrate methane production with the growing of both food and fuel crops provide another example.

In Murray Bookchin's words: "We should always have a diversified mosaic of energy sources — utilising, as it were, all of the forces of Nature so that they interplay with our lives. In this way, we can develop a more respectful — even reverential — attitude towards Nature."

8 Renewable Energy in a Decentralised Society

So far, I have concentrated on the use of natural, renewable energy sources to supply the needs of households, or small communities. But as the Annual Abstract of Statistics makes clear, domestic energy use — though the second largest single category of consumption — is only about 25 per cent of all the energy consumed in the country as a whole.

The biggest consumer of all is industry, with 42 per cent share (710×10^9 kWh), of which the steel industry alone accounts for a massive 11 per cent (185×10^9 kWh). The next biggest consumer, apart from the domestic user, is road transport, which absorbs some 16 per cent (274×10^9 kWh) — almost entirely in the form of oil. Compared to road transport (i.e. the motor car) other transport modes consume insignificant amounts of our energy. The railways account for about one per cent, water transport consumes about 0.7 per cent (11.5×10^9 kWh), and even air transport uses only 3 per cent (52.4×10^9 kWh) — though the low figure for aircraft fuel consumption is more a reflection of the comparatively tiny number of people who travel by air than of any economy in our gasoline-gulping big jets. Other energy users are "public services", with a 6 per cent share (102×10^9 kWh) and agriculture, which accounts for about 1½ per cent (25×10^9 kWh) of the total. The remaining five per cent or so (86×10^9 kWh) is classed in official statistics as "miscellaneous".

We have already seen how in the domestic sphere, those energy requirements that remain, after we have eliminated the profligate waste that prevails at present, could easily be catered for by the intelligent application of various devices for capturing the energy of the sun, wind, water and plants.

But to what extent can these renewable sources supply our industries and run our transport systems, our agriculture and our public services?

The logical approach to this question is very similar to the

91

approach to domestic energy consumption adopted earlier. To begin with, is the energy consumed in these sectors at present being applied to useful purpose? Or is it being wasted as extravagantly as energy in the domestic sphere?

In the case of industry, the answer to the latter question must surely be an emphatic *yes*. Not so much because industry is, from an internal economic point of view, any more inefficient in the way it uses its fuel inputs than the other sectors of the energy economy: but because the products of industry *themselves* are extravagantly wasteful of energy. These products are wasteful in several major ways.

Firstly, many of the products are totally unnecessary — the most obvious example of such redundancy being the superfluous packaging with which goods nowadays are universally surrounded. Secondly, the products have deliberately shortened lifetimes or built-in obsolescence so that they fail far earlier than they would need to do if they were designed properly, and have to be replaced more frequently — which requires an amount of energy equal to the original input.

Thirdly, the products themselves are made from high-energy materials — that is, materials which themselves require great amounts of energy in refining and processing.

Aluminium drink cans are an oft-quoted example: an aluminium beer can uses many times more energy in its manufacture than an equivalent glass container; and the glass container can be reused several hundred times, whereas the aluminium can is useless after it has been ripped open and has to be thrown away. The motor car is perhaps the best overall illustration of the profligacy of our industrial energy wastefulness. Individual motor cars, at least in the numbers in which they are produced nowadays, are largely unnecessary — both in the sense that peoples' normal desire to be able to travel could easily be met by having a greatly-expanded public transport network, supplemented perhaps by a fleet of municipally-controlled car and van hire services for those occasions when a car is necessary; and in the sense that a vast number of daily journeys would be totally superfluous if society were decentralised and people lived within walking distance of their friends and places of work.

Motor cars are also the most blatant example of built-in obsolescence. The cars that are being sold today are not in any significant way more advanced than those being sold 30 years ago. And there is no reason whatsoever why a car should not last for 30 or even 50 years, or more, without the need for more than routine maintenance. On the reasonable assumption that it takes no more energy to manufacture a long lasting car than it does to manufacture a short life car, the energy "consumed" by one 50-year car is only one-seventh of the energy required to build seven seven-year cars.

Moreover, if the materials used to make such long life cars, together with buses and lorries, (in the relatively small numbers that would be needed in a low-energy society) were to be chosen with due regard for the energy cost of refining and processing them, the amounts of energy consumed in transportation could be reduced by a further significant amount.

Houses are another example of wastefulness, not only in their day-to-day "running" consumption of energy, as we have seen, but also in the high-energy content of the materials used to make them. Andrew MacKillop has shown that a typical three-bedroom semi-detached house built to Parker-Morris standards uses something like 53,000 kWh of energy in its construction.[57] But by employing low-energy materials, such as blocks made from 10 per cent cement and 90 per cent soil, stone, and local wood supplies, the energy cost can be cut to 19,000 kWh. Moreover, by substituting rammed-earth construction for the soil-cement blocks, the energy input required can be sliced to about 5,000 kWh — i.e. by a factor of 10 overall.

Another highly important aspect of our present industrial consumption of energy is our "need" to produce vast quantities of goods for export, so that we can afford to pay for imported food. But quite apart from the fact that world food supplies are now getting so tight that few countries will be willing to supply us with the food they need for their own home consumption, (and even these will only be willing to do so at prices which we cannot hope to afford), we are also faced with a situation where most of the developed world, and an

93

increasing part of the "underdeveloped" world, no longer either needs or wants our "high technology" products.

Sooner or later, we will have to face the fact that the myth of Britain as "a trading nation" — a sceptred isle which thrives by exploiting its native skills in the manufacture of sophisticated finished goods in return for imports of food and raw materials — is as good as dead. Sooner or later, and it had better be sooner, we must begin to concentrate our resources on increasing our domestic food production. The only way *that* can be done is to have more people working on the land, so that its productivity per acre can be raised without the use of artificial fertilisers which, even in direct comparison, yield no better tonnages per acre than organic farming methods, and which eventually leave the soil exhausted, and pollute our rivers.

The whole question of food production is one which could fill several books, however, and I cannot do justice to the issue in a few paragraphs. But such a "back-to-the-land" policy would not only mean more food, so there would be less need for industrial production for export, with its associated high energy consumption; it would necessarily imply a considerable de-centralisation of population from the overcrowded cities and a more even dispersal of people throughout the country, which in turn would mean a far smaller demand for transport.

A simultaneous expansion in the use of telecommunications could avoid the possibility that such a decentralised Britain would degenerate into a series of inward-looking, parochial communities that existed in previous, pre-electronic eras. Equally importantly, telecommunications could *further* reduce the demand for transportation.

One important example of a telecommunications technique that can make possible significant reductions in our use of transport is facsimile picture transmission — used for decades to transmit news photographs across the world and recently adapted to the transmission of ordinary letters. Even now, facsimile letter transmission systems are available on the market (e.g. the Kalle Infotec 6,000 facsimile transceiver) which will transmit the contents of an A4 letter over any distance, using standard telephone lines, in 35 seconds. To send such a letter within Britain, at normal telephone call rates, the cost would be 1p at standard rate for up to 35 miles, and 4p at standard rate for over 35 miles. At cheap rates, after 6 pm and at weekends, the cost would always be less than 1p. But even at 4p, the cost compares pretty favourably with the current 4½p first class letter rate, especially considering the virtual instantaneity of transmission time (mid-1974 prices).

Further developments in the efficiency of document scanning promise to reduce the amount of redundant information and to decrease still further the time needed to transmit a letter. Such advances, coupled with the trend towards reduced trunk call charges at off peak times, seem bound to make facsimile transmission competitive with letter post, even allowing for the cost of the machines required. Indeed the only reason why facsimile transceivers are not now as common as office photocopiers is that manufacturers have not yet been able to agree among themselves about a standard transmission format that everyone can use. And a facsimile transmission system which isn't the same for everyone is worse than no facsimile system at all. (Another problem has been lack of switching capacity at Post Office exchanges, but that is slowly being remedied.) Considered purely from an energy standpoint, the facsimile transmission of one page containing about 20,000 bits (say 500 words), as Tribus and McIrvine point out[58] consumes about 20,000 joules of energy — that is, about 0.006 kWh. Which is insignificant compared to the energy expended by the sender of a letter in walking to the post box, or the Postman's energy in walking to the recipient's door — not to mention the energy expended per letter in handling, sorting and transport between the sender's and recipient's Post Offices.

The kind of industrial base needed to support such a small-scale, decentralised society has been outlined by Murray Bookchin in his classic essay "Towards a Liberatory Technology."[59] Bookchin's analysis of how even the steel industry, normally thought of as essentially an ultra-large scale, ultra-high capital, ultra-centralised operation, could be scaled down to a "human scale" is, I think, too important to paraphrase, and worth quoting in full:

"A fascinating breakthrough," he writes, "has been achieved in reducing the size of continuous hot-strip steel rolling mills. A typical mill of this kind is one of the largest and costliest facilities in modern industry. It may be regarded as a single machine, nearly a half mile in length, capable of reducing a ten-ton slab of steel about six inches thick and 50 inches wide to a thin strip of sheet metal, a tenth or a twelfth of an inch thick. A hot-strip mill runs the steel slab through scale-breaker stands, roughing stands with huge vertical rollers, and a series of finishing stands. The entire installation, including heating furnaces, coilers, long roller tables, and buildings, may cost in excess of 50 million dollars and occupy 50 acres. It produces 300 tons of steel sheet an hour. To be used efficiently, a continuous hot-strip mill must be operated together with large batteries of coke ovens, open-hearth furnaces, blooming mills, etc. These facilities, in conjunction with hot and cold rolling mills, may cover several square miles. It is a modern steel complex, geared to a national division of labour, to highly concentrated sources of raw materials (located at a great distance from the complex) and geared toward large national and international markets. Even if totally automated, its operating needs and management far transcend the capabilities of a small, decentralized community. The type of administration it requires is essentially national in scope. Its economic weight, in effect, is thrown in support of centralistic institutions.

"Fortunately, we now have a number of alternatives — in many respects, more efficient alternatives — to the modern steel complex. We can replace blast and open-hearth furnaces with electric furnaces. These are generally quite small and produce excellent pig iron and steel; they operate not only with coke as a reducing agent, but also with anthracite coal, charcoal, and even lignite. Or we can choose the HyL process, a batch process in which high-grade ores or concentrates are reduced to sponge iron by means of natural gas. Or we can turn to the Wiberg process in which reduction is achieved by the use of carbon monoxide and a little hydrogen. In any case, we can eliminate the need for coke ovens, blast furnaces, open hearth furnaces, and possibly even solid reducing agents.

"But the most important step in the direction of scaling down the size of the steel complex to community dimensions is the development of the planetary mill by T. Sendzimir. The planetary mill reduces the typical continuous hot-strip mill to a single planetary stand and a light finishing stand. Hot steel slabs, 2¼ inches thick, pass through two small pairs of heated feed rolls and a set of work rolls, mounted in two circular cages, which also contain two back-up rolls. By operating the cages and back-up rolls at different rotational speeds, the work rolls are made to turn in two directions. This gives the steel slab a terrific mauling and reduces it to a thickness of only one-tenth of an inch. Sendzimir's technique can be regarded as a stroke of engineering genius: the small work rolls, turning on the two circular cages, are given a force that can only be achieved by four huge roughing stands and six finishing stands in a continuous hot-strip mill. What this means is that the rolling of hot steel slabs requires a much smaller operational area than that occupied by a continuous hot-strip mill. With continuous casting, moreover, we can produce steel slabs without the need for large, costly slabbing mills. Taken altogether: Several electric furnaces, the use of continuous casting, a planetary mill, and a small, continuous cold-reducing mill, occupying little more than an acre or two, would be fully capable of meeting the steel needs of a moderate-sized community. This small, highly sophisticated complex would produce an extremely high grade of steel and involve substantially lower heat costs and scale losses. Without automation, it would still require fewer men to operate, even if we account for its lower output level, than a conventional steel complex. It could reduce lower grade ores more efficiently and with less difficulty. And finally, since the planetary mill produces a shiny and clean strip for cold rolling merely with high-pressure water, it eliminates acid-pickling and the need to dispose of waste-pickling liquor — a major source of stream pollution caused by conventional steel plants.

"The complex I have described is not designed to meet the needs of a national market of the kind that exists in the United States today. It is suited for meeting the steel requirements of small- or moderate-sized communities and industrially underdeveloped countries. Most electric furnaces produce about 100 to 250 tons of molten iron a day, compared with new large blast furnaces that produce 3,000 tons daily. A planetary mill can roll only a hundred tons of steel strip an hour, roughly a third of the output of a continuous hot-strip mill. Yet the very productive scale of our hypothetical steel complex constitutes one of its most desirable features. Owing to the more durable steel produced by our complex, the community's need to continually replenish its steel products is appreciably reduced. Since the complex requires ore, fuel, and reducing agents in only small batches, many communities can rely on local resources for their raw materials, conserving the more concentrated resources of centrally located sources of supply, strengthening the independence of the community itself vis-a-vis the traditional centralized economy, and reducing the expense of transportation. What may seem to be a costly, inefficient duplication of effort that could be solved by a few centralized steel complex would prove, in the long run, to be more efficient as well as socially more desirable."

Bookchin also explores the potential of *multi-purpose* manufacturing machinery — a concept diametrically opposed to the trend of centralised mass-production industry over the past few decades, which has been towards ever-increasing *specialisation* of tools and machinery. He cites the example of a new horizontal boring machine that is capable of drilling holes "smaller than a needle's eye or larger than a man's fist", accurate to one ten-thousandth of an inch. Such machines, and others like them, could make it possible "to produce a large variety of products in a single plant. A small or medium sized community using multi-purpose machines could satisfy many of its limited industrial needs without being burdened with under-used industrial facilities.

"The community's economy would be more compact and versatile, more rounded and self-contained, than anything we find in the communities of the industrially advanced countries."

"I do not claim," he admits, "that *all* of man's economic activities can be completely decentralised, but the majority can surely be scaled down to human and communitarian dimensions. This much is certain: we can shift the center of economic power from national to local scale and from centralised, bureaucratic forms to popular, local assemblies."

As an exercise, it is interesting to put some figures on all these speculations to see just what a low energy society's needs might actually be.

We have already dealt with domestic energy consumption (420 x 10^9 kWh a year, or 25 per cent of total consumption) and shown that by adopting various techniques and cutting waste, demand can fairly easily be cut by nearly 50 per cent from 23,000 kWh per household per year to something like 12,800 kWh, and brought into line with the available supply of renewable energy sources.

Let's now look at the other energy-consuming sectors of society. Industry is the most important sector requiring examination, since industry, including iron and steelmaking, consumes some 42 per cent of all our energy.

As I have suggested above, British industry could cut its energy consumption by an enormous amount if it: (a) adopted energy-conserving practices such as the use of total energy schemes, better insulation, more efficient ventilation and so on; (b) abolished the use of all unnecessary packaging, and the production of unnecessary goods (not to be confused with *luxuries,* at least some of which are, in a sense, a necessity in any society); (c) made goods to last as long as possible and (d) made goods from low energy materials.

A significant illustration of British industry's ability to save energy was, by the way, provided during the 1974 miners' strike, when most firms maintained almost full output despite massive power restrictions, and the overall effect on production was far smaller than the "experts" had predicted.

It seems reasonable to assume that just by using fuel and electricity economically, applying the same-techniques as I have suggested for the domestic sector, British industry could cut its energy use by 50 per cent, and still maintain the same level of output. By cutting back on the use of packaging, and on the production of trivia (but not some luxuries), let's say 10 per cent more could be saved. Which brings industrial energy consumption down to, say 40 per cent of its present 710 x

10^9 kWh, or to 284×10^9 kWh. If we were to make goods last for 50 years instead of the usual maximum of 10 years, then it should be possible to slice industrial energy needs by at least a factor of five — but let's assume that consumption might only decrease by a factor of three, to about 95×10^9 kWh a year. And by substituting low energy materials, even this figure could probably be cut by, say one third, to give a total net *necessary* demand of 62×10^9 kWh a year. Let us assume, however, that the net necessary industrial demand is somewhat higher — say, for convenience, 10 per cent of the current gross demand, or 72×10^9 kWh a year.

To turn now to transport, the third-largest item in the country's energy budget: Transport can conveniently be divided into two categories — passenger transport, and goods transport. For the former, the most common unit of traffic measurement is the *passenger mile,* one passenger mile being expended when one passenger travels for one mile. For the latter, traffic is measured in *ton miles,* one ton mile being expended when one ton of freight is carried over a distance of one mile.

In Britain in 1972, the total number of passenger miles travelled was 278.8 thousand million, of which air travel accounted for 0.45 per cent, rail travel for 7.5 per cent, public road transport for 12.3 per cent, and private road vehicles for a massive 80 per cent. Road transport, both public and private, accounts therefore for 92.3 per cent of all passenger travel.

In the case of goods transport, the overall total of 80.6 thousand million ton miles expended in 1972 was made up as follows: road, 64 per cent; rail, 17.6 per cent; coastal shipping, 16.2 per cent; inland waterways, 0.124 per cent; and pipelines, 2.36 per cent.

The fact which leaps out from these statistics — that road transport, which nowadays is almost exclusively powered by the internal combustion engine, is far and away the most widely used mode of transport — is parallelled by the fact that the internal combustion engine is one of the least efficient of all devices for converting energy. The engine itself has an efficiency of only about 25 per cent, but when the overall efficiency of oil-powered transport is considered, taking into consideration all the losses incurred in drilling and refining the petroleum, and in transporting it from refineries to filling stations, plus the mechanical losses in the vehicle's gearing and transmission system, the net efficiency of the average automobile falls to something like 12 per cent.[60]

In terms of the amount of energy required per unit of traffic, a car needs between 1.2 and 2.4 kWh per passenger mile (depending on the number of occupants). A commuter train, on the other hand, with only 50 per cent of its seats filled, needs only about 0.4 kWh per passenger mile, while lightweight trains and buses consume even less — about 0.3 kWh per passenger mile. A juggernaut lorry is reckoned to consume one US gallon of fuel for every 50 ton miles of freight carried, which is equivalent to about 0.8 kWh per ton mile. But rail, waterway and pipeline transport of freight is much more economical — about 0.2 kWh per ton mile. Aircraft are even more extravagant than motor vehicles: a Boeing 747 requires something like 1.75 kWh of energy per passenger mile — nearly five times as much as a commuter train.[61]

All these facts go a long way towards explaining why, as quoted earlier, road transport consumes a lion-sized 16 per cent share of the UK's energy, compared to 3 per cent for air transport, one per cent for the railways, and 0.7 per cent for water transport.

The total, about 21 per cent of the country's energy budget, is, however, a considerable *under-estimate* of the actual energy resources absorbed by transport, because it does not take into account the "secondary" energy inputs necessary to maintain the transport system — the energy needed to make asphalt for roads, concrete motorways bridges, motorway lamp standards, and so on. Grimmer and Luszczynski[62] have calculated that in the USA, such secondary inputs increase transport's share of national expenditure by about a third. There is no reason to assume that a similar factor is not applicable in this country, which would raise transport's share of our energy to about 28 per cent.

But for simplicity, let's stick to the official overall figure of 21 per cent, and 16 per cent for road transport.

It is obvious that by substituting more efficient modes of

transport for the motor car, considerable savings in energy could be effected. Let's suppose that we were to cause a radical shift in the emphasis of our transport system by switching 90 per cent of all road passenger traffic in private cars to other, more efficient, systems — let's say to the railways, to buses, and to inland waterways. Such a situation would not be radically different to that which prevailed in this country before the post-war motor vehicle boom.

Private car traffic would now account for some 22 thousand million passenger miles, instead of 222 thousand million at present. At 1.2 kWh per passenger mile, the energy consumed would be 26.5×10^9 kWh per year. The remaining 257 thousand million passenger miles would be handled by bus, rail and waterway, at an energy cost of, say, 0.3 kWh per passenger mile, or 78×10^9 kWh a year in toto.

Similarly, goods traffic travelling by "juggernaut" lorry could be cut to 10 per cent of its present level, to about 5.14 thousand million ton miles. At 0.8 kWh per ton mile, this reduced traffic would burn up about 4.1×10^9 kWh a year. The remaining goods traffic, some 75 thousand million ton miles, if sent instead by rail, barge or pipeline at an energy cost of 0.2 kWh per ton mile, would require some 15×10^9 kWh of energy.

Adding all these figures up, we get a total, for passenger and goods transport, of 123×10^9 kWh a year.

Adding on, without reduction, our present air transport consumption of 52.4 kWh a year (on the highly pessimistic assumption that *none* of our present air passenger or cargo traffic could be transferred to rail, barge or lorry) the grand total of energy consumption works out at 176×10^9 kWh a year, or almost exactly half the present level of 354×10^9 kWh a year — for, let us remember, *exactly the same* annual traffic in passenger and ton miles.

Our postulated decentralised society, however, where nearly everyone would work close to home, where production would be mainly for local, not national or international markets, and in which telecommunications would eliminate the need for a great deal of routine travel, would be highly unlikely to require anything like this amount of traffic.

Let's assume that travel could be cut to, say, 50 per cent of its present intensity (though a far lower figure seems more likely). The energy consumption would then be roughly half the figure just calculated, or 88 x 10^9 kWh a year.

Turning briefly towards the other remaining sectors of the energy economy — namely "public services", "miscellaneous", and "agriculture", which consume 6 per cent, 5 per cent and 1.4 per cent respectively — it seems equally possible that energy consumption here too could be trimmed very considerably.

In the public services sector, for instance, the energy consumed in activities like refuse collection and sewage pumping would be largely unnecessary in a society in which organic waste products were regarded as a valuable resource for use in making methane and fertiliser, and in which "scrap" and "rubbish" would be largely absent because of the much longer lifetime of products.

The energy consumed in the agricultural sector today is mainly the result of the increasing mechanisation inherent in large-scale, capital-intensive farming. But to achieve the yields *per acre* (rather than *per man hour)* needed to make Britain self-sufficient in food, a labour-intensive approach will have to take the place of our present energy- and capital- intensive methods.

As for the "miscellaneous" sector, there is probably as much slack present here as there seems to be everywhere else.

But instead of attempting to put a value on the reduced energy demands that might be needed by these rather vaguely defined sectors in a more rational energy economy, I propose to assume that consumption in these areas stays the *same* as at present — at 213 x 10^9 kWh a year. In any case, even the achievement of dramatic savings in these fields would only reduce the total overall consumption by a fairly small percentage, since these sectors account for a much smaller proportion of energy expenditure than the Domestic, Industrial and Transport sectors.

Adding together, then, our "reduced" energy demands for all the various sectors — Domestic at 210 x 10^9 kWh a year; Industry at 72 x 10^9 kWh a year; Transport at 88 x 10^9 kWh

103

a year; and "public services", "agriculture" and "miscellane-ous" at 213×10^9 kWh a year — the grand total comes to 583×10^9 kWh a year, which is about *one third* of our present total energy consumption of 1700×10^9 kWh a year.

The question of how this reduced demand can be met in the *domestic* sphere by the harnessing of renewable energy sources has been dealt with in some detail in earlier chapters. Individually tailored systems, using a combination of solar collectors, wind generators, heat pumps and plant-powered "total energy" schemes, seem more than capable of supplying all reasonable home energy needs.

As for the energy economy as a *whole*, it was deduced in chapter 2 that by burning high-efficiency plants like sugar beet in a "total energy" heat engine it should be possible — at least in theory — to meet all the present energy requirements of the UK (*including* a basic 2,400 calories of food energy per person), by growing such plants on one third of the UK's arable land. To meet our *reduced* energy demands, therefore, only about one ninth of the country's arable acreage would be required. Bearing these figures in mind, let's look at individual energy sectors more closely.

The provision of *transport* using renewable energy resources presents a few problems. It is difficult to power a bus or train in any reliable fashion using wind, water or direct solar power, though methane-fuelled transport is an entirely feasible proposition — as many enterprising motorists demonstrated during the last war when petrol was scarce. But methane, as pointed out in chapter 2, presents storage problems which make it difficult to provide methane-powered vehicles with enough gas for more than about 50 miles of motoring (for this reason, it has been half-seriously suggested that 50 miles should be the average spacing between communities in a decentralised society!).

A bigger disadvantage with methane is that, if it is burned in a conventional internal combustion engine, its energy is being utilised with at best only 25 per cent efficiency. Moreover, in a moving vehicle there is very little one can do with the "waste" heat rejected by such an engine except to pump it out through the radiator into the surrounding environment: mobile "total energy" systems are somewhat impractical. Rail travel, at least with electric locomotives, is a different matter however, because electric motors are very efficient (about 90 per cent), and losses in transmission along the overhead wires are fairly small. Electricity for powering trains in a future decentralised society could easily be generated at a large number of methane-powered total energy power stations, situated perhaps at each railway station along the track and supplying their "waste" heat to homes in the towns and villages served by the railway. An overall efficiency of 75 per cent should be quite feasible with such an arrangement.

Battery-driven electric vehicles, for similar reasons, would be the most efficient way in which a decentralised society could meet its limited need for individual, door-to-door transport. Electric cars themselves have an efficiency of about 65 per cent[63], and assuming that such cars are supplied with power from an 80 per cent efficient total energy system, an overall efficiency of 50 per cent is attainable, which compares very favourably with the petrol-powered car's 12 per cent.

Energy for industry could also rely heavily on methane — both for direct burning in ovens, furnaces and the like, and for use in electricity-generating total energy schemes. Even now, industry is the largest single user of North Sea Gas, which is essentially methane.

Other industrial energy inputs could include water power, which would imply a return to the old practice of siting industries on rivers; wind power, on a scale larger than the domestic user would consider practical; and perhaps geothermal and tidal power in suitable locations.

Andrew MacKillop's calculations, referred to in chapter 6, imply that there is some 35×10^9 kWh of untapped water power in Britain, a great deal of which could undoubtedly be harnessed by local industries. Large scale wind power schemes are also capable of generating very considerable quantities of energy — witness the famous 175 ft diameter, 1250 kW rating machine installed at Grandpa's Knob in Vermont during the last war.

Interest in the potential of wind power in the UK has recently been reviving. Dr. Arthur Bruckner of Louisiana State

University, visiting senior research fellow at City University, London, has suggested in *New Scientist*[64] that a network of large (up to 2,000 kW rating) aerogenerators (linked to a pumped storage reservoir scheme which would supply power during periods of low wind) could be competitive in cost-per-kWh terms with present-day fossil fuel power stations. Pilot studies conducted by Dr. Bruckner at City University using the power needs of Ulster as an example, have indicated that 20 per cent of the province's present electricity demand could be met with as few as 700 aerogenerators linked to a pumped storage scheme, even if the "load factor" (the percentage of generated power actually consumed) was quite low. As Dr. Bruckner puts it, "Wind energy is an ideal, low-cost fuel, providing a non-polluting energy supply on a self-financing basis" (i.e. without the need for fuel imports from abroad.) Referring to the Electrical Research Association's work on aerogenerators nearly 20 years ago, he points out that "from the work it did in the 1950's, Britain leads the world in wind energy research, in aerogeneration and hill-wind dynamics."

Dr. Bruckner believes that we should therefore begin to develop wind energy sources as rapidly as possible, taking advantage of what he discreetly refers to as "under-utilised design resources within the aircraft industry." Translated to the aerospace domain, the slogan "swords into ploughshares" could well become "Concordes into windmills".

Geothermal power is another possible industrial source of renewable energy. New techniques being developed at Los Alamos scientific laboratory may bring this power source within the reach of communities which do not happen to be near hot springs or geysers. According to Nigel Hawkes[65] these techniques involve drilling a narrow hole several thousand feet down into the earth's crust, until it reaches layers of hot rock which have a temperature of around 200 degrees centigrade. Water would then be piped down the hole, turned into steam by the hot rock, and brought back to the surface by a second pipe, where it could be used to drive a steam turbine directly. Each hole would cost about **$1** million, however, and the steam could be drawn off for only something like 20 years before the rocks would cool to a temperature low enough to prevent the generation of steam, so such a power source would not really be strictly "renewable" — even though one could always drill fresh holes in different locations, and though one would be unlikely to be able to deplete more than a tiny fraction of the earth's stored geothermal heat.

The harnessing of tidal power is another proposal that has been given serious consideration in Britain recently. Dr. Tom Shaw of Bristol University has put forward[66] a grandiose £1,000 million project to tame the tides of the Severn estuary by erecting a huge barrage across the Bristol channel, which would not only generate electricity with the ebb and flow tides, but would be designed to act as a pumped storage power reservoir.

But tidal power can be harnessed on a far smaller scale than this. Apart from the famous Rance estuary scheme in France, which supplies some 60 million kWh a year, small tidal mills have, as Stefan Szczelkun points out[67], been in existence since the 11th century, and even in Britain there are a few remaining — for example at Woodbridge in Suffolk and Carew in Pembroke.

The Bristol Channel mega-project — like some of the plans for geothermal power stations and designs for monster windmills — illustrates once again the disease of creeping gigantism which seems to threaten all attempts to harness renewable energy sources. Even if large-scale energy projects are, in theory, worker- and community-controlled, they carry the built-in danger that their very size will make them less amenable to control by the individuals they are meant to serve. The administration of such projects runs the constant risk of becoming bureaucratised and insensitive to the real needs of the communities which use the energy they generate.

Some degree of centralisation, as Bookchin concedes, is probably inevitable, but it should be regarded as a necessary evil, to be tolerated as seldom as possible, and only under strict conditions. To return to a society based on the traditional energy sources — Sun, wind, water and plants — will create formidable withdrawal symptoms in a civilisation which, like ours, has become almost hopelessly hooked on the large amounts of energy provided by the "quick fix" of fossil fuels. This dependence has now become so complete that our rulers, like desperate heroin addicts, are willing to risk anything — even the genetic future of our descendants, in the case of nuclear power — in order to ensure that their mainline supply does not dry up.

Many people, deluded by the seeming solidity of our civilisation into thinking that fossil fuels and nuclear power will last for ever, will see any "back to the Sun" suggestion as a decidedly retrograde step, because they confuse "progress" with the endless pursuit of more and more of what we have now. "But if one is standing on the edge of a precipice," as someone sagely remarked, "the only sensible form of progress is to step backwards." Far from being retrograde, such a "step backwards" would in fact imply a tremendous advance in the net sum of human happiness. But do we have the vision to take that step backwards from the juvenile technocratic fantasies of the "space age" and go forward in a different direction to create a warmer, more natural, more truly human society — a society like that dreamed of by Cobbett, a society blessed by "flocks, herds, corn, wine and oil; smiling land; a rejoicing people; abundance for the body and gladness of the heart"?[68]

If we do not, the alternative is phrased starkly by Aldous Huxley:

"Unless we choose to decentralise and to use applied science, not as the end to which human beings are to be made the means, but as the means to producing a race of free individuals, we have only two alternatives to choose from: either a number of national, militarised totalitarianisms, having as their root the terror of the atomic bomb and as their consequence the destruction of civilisation (or, if the warfare is limited, the perpetuation of militarism); or else one supra-national totalitarianism, called into existence by the social chaos resulting from rapid technological progress in general and the atom revolution in particular, and developing, under the need for efficiency and stability, into the welfare tyranny of Utopia".[69]

The choice before us is presented in similar terms by G. K.

Chesterton in his preface to Cobbett's *Cottage Economy*. "We must go back to freedom or forward to slavery. The free man of England, where he still exists, will doubtless find it a colossal enterprise to unwind the coil of three centuries. It is very right that he should consider the danger and pain and heart-rending complication involved in unwinding that coil. But it is also proper that he should consider the alternative: and the alternative is being strangled."[70]

References

1. Kropotkın, P. *Fields, Factories and Workshops.* These quotations are taken from the long out-of-print 1901 Edition, published in London by Swan Sorranschein. But this great classic is, thankfully, available again in 1974, George Allen & Unwin, edited by Colin Ward.

2. See King Hubbert, M. "The Energy Resources of the Earth". *Scientific American.* September 1971.

3. See *Scientific American* September 1971.

4. King Hubbert, M. *Ibid.*

5. See *Sunday Times,* December 30, 1973.

6. Gofman, J. W. and Tamplin, A. R. *Poisoned Power* Chatto & Windus, 1971.

7. Gofman, J.W.. "Is Nuclear Fission Acceptable?" *Futures,* September 1972

8. Science is working on the problem, though. Peter Glaser, vice president of Arthur D Little Inc., put forward in 1970 his idea for a huge solar power station in orbit around the earth, which would transmit its power back to ground level by microwave radio beam. Glaser suggests that an enormous panel of solar cells, 25 square miles in area, could generate electricity to power a battery of microwave generators. These would feed their power to a one-kilometre wide transmitting antenna, which would be focussed on a 7km-wide receiving antenna on Earth.

The microwave beam would pass straight through any clouds with little absorption, and the receiving antenna would be connected to the normal electricity grid system, to which it would supply something like 10,000 megawatts of power. The advantages, according to Glaser, are that the "Satellite Solar Power Station" (SSPS) would be able to operate at full efficiency for nearly 24 hours a day, unlike terrestrial solar power stations which cannot operate at night or when there is cloud. The disadvantages, which Glaser plays down, are that the capital cost per kilowatt of each SSPS would be at least twice as great as that of a conventional power station; that such stations, by capturing solar energy, which would not normally be received by the Earth, could significantly alter the planet's "heat balance"; that the microwave beam, even if correctly pointed, is of sufficient intensity to cause concern about its possible effect on living tissue; and that the microwave beam could accidentally (or deliberately for military purposes) be deflected from its focussing point at the receiving antenna to a location where its radiation could harm human beings. But the biggest objection to the scheme is that, like all the other high-technology megaprojects, such stations could only be constructed and operated by the richest corporations or nations of the world, in whose already over-powerful hands they would concentrate even more control.

9. Patterson, Walter. *Nuclear Reactors.* Earth Island, 1973.

10. *Annual Abstract of Statistics.* HMSO, 1973.

11. Diamant, R M *Total Energy* 1970

12. Bell, Boulter, Dunlop & Keiller, *Methane, Fuel of the Future,* Andrew Singer Publisher, The Mill House, Coleshill, nr. Highworth, Wiltshire.

13. Marsh, G J W, *Water for six.* Marley Plumbing Technical Publication No. 5., Lenham, Kent. 1971.

14. National Building Agency Report *The Economic and Environmental Benefits of Improved Thermal Insulation.* 1967 (Rev 1969).

15. See Smith, Gerry E. *Economics of Solar Collectors, Heat Pumps and Wind Generators,* Working Paper No. 3, University of Cambridge Department of Architecture, Technical Research Division, 1, Scroope Terrace, Cambridge.

16. Bell, Boulter, Dunlop & Keiller, *ibid.*

17. A commercial version of the Hay Box, using modern synthetic insulating materials instead of hay, recently appeared on the British market. Called "Tabitha's Fireless Cooker". It is claimed to be capable of saving between 50 and 70 per cent of the energy used in conventional cooking, and is available from Low Impact Technology Limited, Wadebridge, Cornwall.

18. See Shelwell-cooper, *Eating Without Heating,* John Gifford, 1943.

19. An ingenious invention which utilises the heat contained in stale air being extracted from a building has been developed by the Swedish engineer Carl Munter. "Munter's Wheel", as it has been christened, enables the heat contained in outgoing stale air to warm up incoming fresh air. The device is marketed in Britain as the Trianco Econovent Heat Recovery Unit, and is obtainable from Trianco, Imber Court, East Moseley, Surrey. At present the unit is designed for industrial purposes and is a little too large for domestic use, but there is no reason why a smaller version should not be produced if there were sufficient demand to justify production.

20. Brinkworth, J.B. *Solar Energy for Man,* Compton Press, 1972.

21. See *Alternative Sources of Energy* No. 11, July 1973. Available from Route 2, Box 90-A, Milaca, MN56353, USA. $5.00 for 6 issues.

22. Even in Britain, the cultivation of crops such as sugar beet seems to be capable of achieving comparable photosynthetic efficiency. According to Ministry of Agriculture Fisheries and Food statistics, for example, the estimated yield per acre of sugar beet in 1971 was 16.6 tons per acre — i.e., 16.6 . 1016 kg per 4046 sq. metres, or about 4 kg per square metre. Burning this weight of organic matter should yield about 25kWh per square metre. Taking Brinkworth's figure of 900kWh as the actual annual insolation in Britain, this means that the photosynthetic

112

efficiency of sugar beet cultivation in the UK works out at just under three per cent.

23. Hall, D.O. paper presented to the inaugural meeting of the UK section of the International Solar Energy Society, January 1974. Reprinted in *Sun at Work in Britain*, No. 1, the magazine of the society's UK branch, available from the Royal Institution, Albemarle Street, London W1X 4BS. £1.00

24. Taylor, T.B. and Humpstone, C.C., *The Restoration of the Earth*. Harper and Row, New York, 1973.

25. Equivalent to some 44,000 kWh per person per year, which is the gross *per capita* consumption of primary fuels. Each person, allowing for all the losses in generation and distribution, actually consumes, on average, about 31,000 kWh annually.

26. Evans, E.N. "Private Generation — basic considerations," *Electrical Times*, June 15, 1972.

27. Hills, Lawrence D. "Fuel Plus Fertility" pamphlet, available free from the Henry Doubleday Research Association, 20 Convent Lane, Bocking, Braintree, Essex.

28. Moorcraft, Colin. "Solar Energy — Part II, Plant Power", *Architectural Design* 1, 1974.

29. Hills, Lawrence D. *op. cit.*

30. Moorcraft, Colin. *op. cit.*

31. Sholto-Douglas, James. *Hydroponics, the Bengal System* Oxford University Press, London. Fourth Edition, 1970.

32. see *Methane Digesters for Fuel, Gas and Fertiliser*, and the stunning *Journal of the New Alchemists*, available from the New Alchemy Institute, PO Box 432, Wood Hole, Massachusetts, USA.

33 Street Farm House, Thames Polytechnic Playing Fields, Kidbrooke Lane, London SE9.

34. Fry, L. John, *Methane Digesters for Fuel, Gas and Fertiliser*, *op. cit.*

35. see Moorcraft, Colin. "Solar Energy in Housing", *Architectural Design* 10, 1973.

36. Moorcraft, Colin, "Solar Energy in Housing", *ibid*.

37. See *Alternative Sources of Energy*, No. 10, March 1973., and *Undercurrents* No. 5, Autumn/Winter 1973.

38. see Moorcraft, Colin, "Solar Energy in Housing", *ibid*.

39. see *Popular Mechanics*, February 1965, *Mother Earth News*, May 1971, and *Undercurrents*, May 1972.

40. Brinkworth, J.B. *ibid*.

41. Vale, Robert. Results of Solar Collector Study, working paper No. 12, University of Cambridge Department of Architecture, Technical Research Division.

42. MacKillop, Andrew. "Living off the Sun", *The Ecologist*, July 1973.

"Energy, Britain's ray of sunshine for Third World".

43. Tucker, Anthony. *The Guardian,* April 15, 1974.
44. Moorcraft, Colin. "Solar Energy in Housing", *ibid.*
45. If we assume (using MKS units — metres, kilograms and seconds) that the velocity of the air in metres-per-second is V, the area of the surface in square metres is A, and the density of the air in kilograms per cubic metre is d, then the total mass M of air molecules hitting the surface per second is equal to the volume of air "sweeping" on to the surface per second, multiplied by its density, i.e. M = Vol. d

(A point "." is used to denote multiplication). But the volume of air swept per second is equal to the air velocity multiplied by the swept area, ie. Vol = V. A, hence M = V. A . d

If we assume, initially, that this mass of air is "stopped dead" when it hits the surface, the amount of energy transferred to the surface per second is equal to the kinetic energy of the moving mass, which by the well-known formula (based on Newton's laws) is equal to ½. M. V^2.

Hence, Energy per second = ½ M . V^2. But M = V . A. d, so

Energy per second = ½ . V. A. d. V^2 = ½ . d . A . V^3

But Energy per second equals *power,* so P = ½ . d . A . V^3

And taking d, the density of air, to be about 1.3 kilograms per cubic metre (the actual density varies somewhat with temperature and pressure) the maximum theoretical amount of power capable of being generated by the given wind flowing across the given surface is equal to

$$P = \frac{0.65 \,.\, A \,.\, V^3}{1,000}$$, where P is in kilowatts.

If the surface is circular, then its area, $A = \frac{3.142 \,.\, D^2}{4}$

where D is its diameter. P then works out at 510 . 10^{-6} . D^2 . V^3

If we take as our wind-blown surface the blades of a windmill, then it is obvious that not all the air molecules flowing past the blades are actually going to hit the blades (though in a well designed windmill one assumes that a large proportion would) and even those that do hit the blades do not lose all their velocity in the process. Aerodynamically shaped windmill blades, by taking advantage of the "lift" available when air flows past an aerofoil surface, are much more effective than flat blades, however. The absolute maximum amount of energy actually capable of being caught by a windmill is reckoned to be some 60 per cent of the theoretical figure given above. And because of losses in friction, power transmission, and so on, a practical figure of something like 35 per cent is more likely.

46. Golding, E.W. *The Generation of Electricity by Wind Power.* Spon, London, 1955.

47. I will be mentioning Conservation Tools and Technology Ltd. quite frequently from now on, in citing examples of actual commercially-available hardware, because CTT is one of the few UK suppliers of such hardware. But of course I cannot vouch for the value-for-money represented by CTT products, though I have no reason to doubt that the company sells anything other than well-made goods at fair prices.

48. Smith, Gerry. *Economics of Solar Collectors, Heat Pumps and Wind Generators, ibid.*

49. see Ott, Dr. John. *Health and Light* from the Devin-Adair Company, One Park Avenue, Old Greenwich, Connecticut, CT 06870, USA

50. Dunlite windmills are now manufactured by Pye of Australia, a subsidiary of the Dutch-based international electrical conglomerate, Philips. This is just one example of the way in which big companies appear to be quietly buying their way into what will probably become known on the Financial pages as the Renewable Energy Market.

51. See for example "The Secenbaugh O_2 Powered Delight" in Mother Earth News No. 20. Plans for this design are available from Jim Secenbaugh, 673 Chimalus Drive, Palo Alto, California 94306.

52. See *Science Dimension*, October 1972. Published by the National Research Council of Canada's Aeronautical Research Establishment.

53. MacKillop, Andrew, "Low Energy Housing, *The Ecologist*, November 1972.

54. Woolston, George, "Peoples Water Power", *Undercurrents* No. 6, March April, 1974.

55. See *Transactions of the Second International Symposium on Molinology*, available from Selskabet Danske Mollers Venner, Brede, Lyngby, Denmark.

56. See Hoda, Mansur. *Impact of Science on Society*, Vol. 23, No. 4, UNESCO, Paris, 1973.

57. MacKillop, Andrew, "Low Energy Housing", *The Ecologist*, November 1972. Somewhat different figures are given by Graham Brown and Pete Stellon in "The Energy Cost of a House", their contribution to a collection of articles on the work of the Rational Technology Unit at the Architectural Association, London. Available from 34—36 Bedford Square, London WC1.

58. Tribus, M. and MacErvine, E.C., "Energy and Information", *Scientific American*, September 1971.

59. Bookchin, Murray, Towards A Liberatory Technology , Ramparts Press, San Francisco, California, and Wildwood House, London.

60. Grimmer and Liczynski "Lost Power", *Environment*, April 1972.

61. *Ibid.*

62. *Ibid.*

63. *Ibid.*

64. Bruckner, Dr. Arthur, *New Scientist,* March 28, 1974.
65. Hawkes, Nigel. *Observer,* March 10, 1974.
66. see *Sunday Times,* March 17 1974.
67. Szczelkun, Stefan. *Survival Scrapbook: Energy.* Unicorn Books, 1974.
68. Cobbett, William. *Cottage Economy.* Cedric Chivers Portway Reprints,
       Bath, 1966.
69. Huxley, Aldous. Introduction to *Brave New World* Chatto & Windus .
70. Chesterton, G. K. Introduction to *Cottage Economy, op cit.*

BIBLIOGRAPHY

The author would like to thank Gerry Foley, George Kasavov and others at the Architectural Association for their help in compiling this bibliography and reference sources.

ALTHOUSE, A. MODERN REFRIGERATION AND AIR-CONDITIONING McGraw Hill, 1968. The most comprehensive book on the subject to date, dealing with every aspect, from simple physics to the finer points of servicing.

AMBROSE, E. HEAT PUMPS AND ELECTRIC HEATING Wiley, 1966. Before you get this book make sure you know something about heat pumps. Not for beginners.

ANDERSON, M.E. QUESTIONS AND ANSWERS: REFRIGERATION McGraw Hill, 1971. Small compact book dealing in the principles of refrigeration. Perfect introduction to the basic theory of heat pumps.

BASSETT, C.D. YOUR OWN WATER POWER PLANT, PART 1, 2 & 3. *MOTHER EARTH NEWS* No. 13, pp. 23-33, Part 4 & 5, No. 14 p. 25—31.

BATE, Harold METHANE GAS PRODUCTION From H. Bate, Penny Rowden, Blackawton, Totnes, Devon. Printed handout. Describes Bate's auto gas converter and the design of a methane digester made from an old hot water cylinder.

A set of 5 cards about methane based on Bate's information are available from Beau Geste Press, Langford Court, 8, Collumpton, Devon. (It is called "manifesto Pamphlet Sitting Dog & Co")

BOHN, Hinrich L. A NEW CLEAN GAS Environment: Vol. 13, No. 10, December 1971 pp 4—9. About utilising agricultural wastes to produce methane.

BRINKWORTH, B.J. SOLAR ENERGY FOR MAN Compton Press, Salisbury 1972. Good introduction to Solar energy, comprehensive coverage, good on theory.

BRS PERFORMANCE TESTS ON SOLAR WATER HEATERS Building Research Station, Overseas Building Notes No. 103. Garston, England 1965.

CAMM, F.J. 'SMALL WIND POWER PLANTS' Newnes *Practical Mechanics,* May 1954 p. 348-350, June 378-380, July 437-438, August 477-478, and September 520-522. Also in the *Second Practical Mechanics Book* Newnes.

Simple methods of constructing small windpower plants from scrap parts and how to rewind dynamos for this purpose.

CHERRY, W.R. CONCLUSIONS AND RECOMMENDATIONS OF THE US SOLAR ENERGY PANEL NSF, RANN & NASA Washington D.C. 1973 These are detailed official feasibility studies.

CHINNERY, D.N.W. SOLAR WATER HEATING IN SOUTH AFRICA. National Building Research Institute, Bulletin 44, South African Council for Scientific and Industrial Research CSIR Report No. 248, Pretoria, South Africa.

CLEWS, Henry ELECTRIC POWER FROM THE WIND, 1973 $1.00 from Solar Wind Co., R.F.D.2, East Holden Maine 04429, USA. Useful little pamphlet, which describes various aspects about installing a small wind generator.

CLOUDBURST — A HANDBOOK OF RURAL SKILLS AND TECHNOLOGY $3.95 from Cloudburst Press Box 79, Brackendale, B.C., Canada. A volume of several reprints from various sources on water wheels etc.

CROWLEY, C.A. 'POWER FROM SMALL STREAMS' PART 1 & 2, Popular Mechanics Sept. & Oct. 1940. Good coverage on how to go about a do-it-yourself water power installation, from water flow survey, to dams and turbines.

COMPOST SCIENCE Rodale Press. Emmaus, Pa 18049, USA. (Patent Office Library). The American recycling magazine. Contains articles on both aerobic and anaerobic (methane) digestion, mainly of town refuse.

DANIELS, Farrington DIRECT USE OF THE SUN'S ENERGY Yale University Press 1964 & 1970. Overview and descriptions of all aspects and uses of solar energy. Good on theory and easy to read.

DEWHURST, J. & McVEIGH, J.C. 'A LOW COST SOLAR HEATER' Heating and Ventilating Engineer, 1968. 41(488), pp. 445—446. Description of flat plate solar collector.

DIAMANT, R. & McGARRY, J. SPACE AND DISTRICT HEATING Iliffe, 1968. Small section on heat pumps — useful piece on how to read heat pump charts.

ELONKA, A. & MINICH, Q. STANDARD REFRIGERATION AND AIR-CONDITIONING, QUESTIONS AND ANSWERS. McGraw-Hill 1961. The book to get if you cannot get Althouse.

FARRALL, A.W. THE SOLAR WATER HEATER Univ. of California, Berkeley, Bulletin 469, June 1969. Detailed info. on a solar heater design.

FLETTNER, Anton THE STORY OF THE ROTOR Willhoft, New York 1926. Good description of the evolution of his rotor and mainly concerned with improving wind propulsion for ships, also a few windmills and describes Savonius rotors. Recommended.

FRY & MERRILL METHANE DIGESTERS FOR FUEL GAS & FERTILIZER. New Alchemist's Newsletter No. 3. 48 page book from N.A. Institute West, 15 Anapamu St., Santa Barbara, California 93101, USA ($3.00).
Very good manual with practically everything you need to know about methane, with designs of simple working models from such components as car inner tubes. Highly recommended. Contains full bibliography.

GAISFORD, Michael MUCK POWER Farmers Weekly May 31, 1974. Excellent hardheaded review of the experiments in farm scale digestion in Britain. Written for farmers and contains no mythology. Good introductory level reading. Useful references.

GARG, H.P. SOLAR WATER HEATER Central Building Research Institute, Roorkee, (UP), India; Building Digest No. 61, November 1968.

GOLDING, E.W. THE GENERATION OF ELECTRICITY BY WIND POWER Spon, London, 1955. Very good coverage of what has been done and a good introduction to the field. The bible of wind power; highly recommended.

GOLUEKE, Clarence COMPOSTING: A STUDY OF THE PROCESS AND ITS PRINCIPLES. Rodale Press Inc.. Book Division, Emmaus, Pa 18049,

1973 $2.95. Covers the history since 1950, the biological principles of the aerobic process, the technology, health aspects and the use of compost. Also contains information on the composting of kitchen and garden by-products.

GOTAAS, Harold COMPOSTING, WHO Monograph No. 31: World Health Organisation 1956 (out of print).

GRAY, BIDDLESTONE & CLARK. REVIEW OF COMPOSTING PART III: PROCESSES AND PRODUCTS Process Biochemistry, October 1973 (Patent Office Library). The history of composting, a review of municipal composting plants, the composting of farm and garden materials. Deals with the final compost product, assessment of quality, decomposition and use. Part IV is not published yet, but will cover the economics, use in agriculture, trace element problems and general potential of composting.

GRAY, SHERMAN & BIDDLESTONE A REVIEW OF COMPOSTING PART· 1 Process Biochemistry, June 1971. (Patent Office Library). A comprehensive review of the art of aerobic composting written by the Compost Studies Group at Birmingham University. Part 1 outlines the need for composting, the biochemistry, time/temperature pattern, humic acid and microbiological aspects, the use of inocula and environmental factors.

GRAY, SHERMAN & BIDDLESTONE. REVIEW OF COMPOSTING PART II: THE PRACTICAL PROCESS. Process Biochemistry, Oct. 1971. (Patent Office Library). The chemical and physical parameters of the practical process. Covers the size of feed material, its C/N ratio and nutrients, moisture content, agitation, aeration and temperature, the oxygen consumption, pathogen destruction, heat production and pH changes. Not written specifically for small scale composting, but very useful.

GRIFFITH, M. SOME ASPECTS OF HEAT PUMP OPERATION IN BRITAIN. IEE Proceedings, Vol. 104, Part A, No. 15, June 1957. Interesting exposition for its time.

HAIMER, L.A. THE CROSS FLOW TURBINE, WATER POWER (London) Jan. 1960. Reprints available from Ossberger Turbinenfabrik, 8832 Weissenburg, Bayern, Germany.

Describes a type of turbine which is being used extensively in small power stations, especially German. Quite good.

HALACY, D.S. THE COMING AGE OF SOLAR ENERGY Harper & Row, New York, 1973. Journalistic introduction to history of solar energy and the present state of the art.

HAMILTON, R.W. Ed. SPACE HEATING WITH SOLAR ENERGY MIT Press 1954. Information on specific behaviour of designs tested in experimental solar houses at MIT and others.

HAMM, Hans W. LOW COST DEVELOPMENT OF SMALL WATER POWER SITES by VITA (Volunteers for International Technical Assistance), 3706 Rhode Island Ave., Mt. Rainier, Maryland 20822, USA. Very good coverage and source book for most aspects of small scale water power installations from water flow measurement, dam building, water wheels and information about water turbines. Recommended.

HAY, Harold 'NEW ROOFS FOR HOT DRY REGIONS' Ekistics No. 183, Feb. 1971, pp. 158—164. Describes Hay's ideas for roof ponds as method of heating and cooling.

HAY, H.R. & YELLOTT, J.I. 'INTERNATIONAL ASPECTS OF AIR CONDITIONING WITH MOVEABLE INSULATION *Solar Energy* Magazine 12. 4pp 427—438, 1969.

HERBERT, William J. 'THE HYDRAULIC RAM PUMP, PERPETUAL MOTION FOR THE HOMESTEAD' *MOTHER EARTH NEWS* No. 22. Description of his installation of a Rife hydraulic engine on his homestead.

HEYWOOD, H. 'SOLAR ENERGY FOR WATER AND SPACE HEATING' Journal of the Institute of Fuel July 1954, pp. 334.
Describes some of his good work in the UK.

HILLS, Lawrence D. SANITATION FOR CONSERVATION Ecologist, November 1972. Outlines the problems of the water borne disposal system, the nutrient value of sewage and the Clivus composter toilet. Contains an analysis of various composts, argues for the use of the Clivus and mentions the Sanivac pneumatic/vacuum system for use in large installations and the Electrolu bath and washing water treatment plant.

HOWARD, Sir Albert AN AGRICULTURAL TESTAMENT Oxford University Press 1940-50. Deals with the relationship of recycling, composting by the Indore method, and soil fertility. Also mentions the relationship between roots, soil microbes and soils.

INTERNATIONAL SOLAR ENERGY SOCIETY UK SECTION. CONFERENCE OF LOW TEMPERATURE THERMAL COLLECTION OF SOLAR ENERGY IN THE UK Central Poly, April 1974. First major attempt to cover British work in solar heating. Includes samples of the work of Heywood, McVeigh, Curtis and Szokalay (though Wallasey School is not included.)

KING, F.H. FARMERS OF FORTY CENTURIES: OR PERMANENT AGRICULTURE IN CHINA, KOREA, AND JAPAN Faber & Faber, 1926. Most interesting account of King's travels in these countries, and the recycling and cultivations that he observed. Though perhaps he did not see the very poor areas, he revealed a nation which had practised energy conservation, careful recycling and the use of human, wind and water energy.

KLEMIN, A. 'THE SAVONIUS WIND ROTOR' Mechanical Engineering Vol. 47, No. 1 Nov. 1925 p. 911. Description of Savonius rotors.

LEFFEL & CO., JAMES, PAMPHLET A, HINTS ON DEVELOPMENT OF SMALL WATER POWER & BULLETIN H—49 Springfield, Ohio, USA. Descriptions of how to go about installing a water turbine and information about the small scale units that they manufacture.

LINDSTROM, Rikart CLIVUS PATENTS:
US 3,136, 608: 9 June 1964
UK 1,126, 520: 5 June 1968

MANNING, P. 'ST. GEORGES SCHOOL, WALLASEY: AN EVOLUTION OF A SOLAR HEATED BUILDING' Architects Journal 25th June 1969. Description of the Wallasey School.

MARIER, Don 'HYDRAULIC RAM' *Alternative Sources of Energy* No. 1. Diagrams and workings of the hydraulic ram. (Also reprinted in *Mother Earth News* No. 22).

MEYER, Hans WIND GENERATORS — HERE'S AN ADVANCED DESIGN YOU CAN BUILD *Popular Science* Nov. 1972 p. 103—105 and p.

142. Plans to build a wind generator from motor bike and car parts, and expanded paper blades.

Michel, Jaques 'CHAFFAGE PAR RAYONNEMENT SOLAIRE' Architecture d'Aujourdhui No. 167, pp. 88—93 (in French) Overview of the Trombe Solar houses in the Pyrenees.

MIT. PROCEEDINGS OF SPACE HEATING SYMPOSIUM. MIT Cambridge, Massachusetts, 1950. Solar heating of housing, collection, storage, energy, transport, architecture, etc.

MOCKMORE C.A. & MERRYFIELD F. 'THE BANKI WATER TURBINE Corrallis, Ore, Oregon State College Engineering Station Bulletin No. 25, Feb. 1949. 40 cents. A translation of a paper of Donat Banki. A highly technical description of this turbine, originally invented by Michell, together with the results of tests.

MOORCRAFT, C. SOLAR ENERGY IN HOUSING Architectural Design, October 1973. Broad overview, includes references to the Wallasey School, Harold Hays Solarchitecture House and the Odeillo Solar Walls.

MOORCRAFT, Colin METHANE POWER Architectural Design, Feb. 1972 p. 131. Good general overview of methane production and use.

PLANT POWER Architectural Design, January and February 1974. Very good investigation and de-bugging of research into using plants as energy converters of solar energy, grown on organic wastes to produce methane, alcohol, etc. Also deals with the uses of combustion and pyrolysis of wood.

MORSE, R.N.
> INSTALLING SOLAR WATER HEATERS CISIRO Circular No. 1, 1959. Melbourne, Australia. Description on how to go about installing a solar water heater.
> SOLAR WATER HEATERS CISIRO Division of Mechanical Engineering, Circular No. 2, Melbourne, Australia. Practical descriptions of how to make solar water heaters.
> SOLAR WATER HEATERS FOR DOMESTIC AND FARM USE. Commonwealth Scientific & Industrial Research Organisation, Engineering Section Report E.D. 5. 1957 Melbourne, Australia.
> 'SOLAR WATER HEATERS — PRINCIPLES OF DESIGN, CONSTRUCTION AND INSTALLATION' Journal Institute of Heating and Ventilating Engineers Jan. 1967. Obtainable from the Building Research Station. Summary of CISIRO Circular No. 2, and describes methods of construction solar water collectors, using conventional materials; good on practical aspects.

PATON, T.A.L. POWER FROM WATER Leonard Hill, London 1961. A concise general survey of hydro electric practice in abridged form. Some chapters on history and water wheels, but mainly straight high-technology stuff.

PUTNAM, P.C. POWER FROM THE WIND Van Nostrand Co., Inc., New York, 1948. Mainly about the development of his wind generator on Grandpa's Knob, Vermont. Some good material on the phenomenon of wind, and measuring windspeeds.

REYNOLDS, John WINDMILLS AND WATERMILLS, Hugh Evelyn, London 1970. Very good on history and development of water power,

121

related to traditional water and windmills, including iron foundry and textile mills. (illustrated)

RICHARDS, S.J. & CHINNERY, D.N.W. A SOLAR WATER HEATER FOR LOW COST HOUSING South African CSIR. Report No. 237, 1967 pp. 1—26. Pretoria, South Africa. Good work on solar water heaters.

RYBXZYNSKI, ORTEGA & ALI. STOP THE FIVE GALLON FLUSH Minimum Cost Housing Group, School of Architecture, McGill University, Montreal, Canada, 1973 $1.75. A survey of alternative waste disposal systems. Lists and classifies 50 toilets. Most of these are poorly designed. This shows the need for a safe, simple cheap and self-contained recycling toilet.

SAVINO, Joseph E.D. WIND ENERGY CONVERSION SYSTEMS: WORKSHOP PROCEEDINGS. Washington 1973, from Joe Savino, Lewis Research Centre, Mailstop 500-201 Cleveland, Ohio 44135, USA. NASA-NSF sponsored conference on wind generation, mainly about large scale machines; not too useful, but is the latest state-of-the-art publication.

SAVONIUS, S.J. 'THE S-ROTOR AND ITS APPLICATIONS' Mechanical Engineering Vol. 53, No. 5, May 1931, p. 335. Description of the Savonius Rotor, and ideas for its use by its inventor.

SCOTT, James C. HEALTH AND AGRICULTURE IN CHINA Faber & Faber 1952. Covers experiments in composing faeces, urine and plant materials by labour intensive methods. Attempts were made to reduce nitrogen loss and the incidence of faecal borne disease.

SHORE, John SANITARY SANITY AA Library 1974. Notes on the historic, social and health aspects of the recycling of human and plant by-products. Includes details of aerobic composter toilet experiments and covers the biological principles of the composting process. Many references.

AEROBIC COMPOSTER TOILETS AA Library 1974. Notes on the design, construction and use of small, family sized composter toilets. A shortened version of Sanitary Sanity.

SINGH, Ram Bux BIO-GAS PLANT AND ITS POTENTIAL 11 page booklet from: the Gobar Gas Research Station, Ajitamal, Etawah (U.P.) India (Free). Discussion of production of methane from animal and vegetable wastes with diagram of a 100 cu ft per day plant and questionnaire for design of a Bio-gas plant.

BIO-GAS PLANT GENERATING METHANE FROM ORGANIC WASTES 67 page book from Gobar Gas Research Station (£2.50). Very good book with several designs for methane digester and tables of different gas outputs from different organic wastes, carbon/nitrogen ratios, building digesters. Applicable to farms. Highly recommended.

SINSON, D.A. & HOAD, T. HOW TO BUILD A SOLAR WATER HEATER Do-it-yourself Leaflet L.4. Feb. 1965. $0.75 Brace Research Institute, Macdonald College of McGill Univ. Ste. Anne de Bellevue, 800, Quebec, Canada. Plans to build a solar water heater.

SPANIDES, A.G. & HATZIKADES, A.D. (Ed.) PROCEEDINGS OF 1961 ATHENS CONFERENCE ON SOLAR & AEOLIAN ENERGY Plenum Press, New York 1964. A few good papers, but mainly a re-run of UN 1961 Rome Conference.

STANFORD RESEARCH INSTITUTE. WORLD SYMPOSIUM ON

APPLIED SOLAR ENERGY. Phoenix, Arizona 1955. Johnson Reprint. Includes details of MIT houses, solar heat pump, some work on biological energy conversion.

STELLON, P. THE HEAT THAT COMES IN FROM THE COLD Undercurrents No. 6. Modest article about the history, theory and practice of the heat pump.

SYSTEME 'D' 14 EOLIENNES (WINDMILLS) No. 9 Societe Parisienne d'Edition, 43 rue de Dunkerque, Paris X^e, France, in French. Describes 14 small do-it-yourself wind power plants for electricity and pumping from bicycle and car parts. Very good. Nice diagrams.

SZOKOLAY, Steven & HOBBS, Raymond 'USING SOLAR ENERGY IN HOUSING' RIBA Journal, April 1973, pp. 177—179. Describes research project, for installing solar heating for housing in the UK. Includes computer study. The research led to installation of solar collectors on a house in Milton Keynes.

TABOR, M. SOLAR ENERGY FOR DEVELOPING REGIONS UNESCO, 1973. SC-73/Conf. 801/2. A concise overview of possible current applications by one of the best helio-technologists.

TECHNICAL RESEARCH ORGANISATIONS

BRACE RESEARCH INSTITUTE, MACDONALD COLLEGE, McGILL UNIVERSITY, STE. ANNE DE BELLEVUE 800, QUEBEC, CANADA. They have 60 or so publications on solar energy, water purification, and windpower, including do-it-yourself plans. Publications list available. Very good work.

COMMONWEALTH SCIENTIFIC AND INDUSTRIAL RESEARCH ORGANISATION (CISIRO) MELBOURNE, AUSTRALIA. They have been doing good work for many years and produced several publications.

COMPLES (CO-OPERATION MEDITARRANEINNE POUR L'ENERGIE SOLAIRE) 32 COURS PIERRE-PUGET, 13006 MARSEILLE, FRANCE.

INTERNATIONAL SOLAR ENERGY SOCIETY, UK SECTION, THE ROYAL INSTITUTION, 21, ALBEMARLE STREET, LONDON W1X 4BS. TEL. 01 493 0669. Membership includes subscription to the Solar Energy Journal, which is best source of continuing work: But society seems to be bowing to industry, especially the oil companies (Solar Power Corporation for instance, a subsidiary of Esso), with very expensive annual individual subscription rates, compared to relatively cheap corporate membership with special perks.

ITDG (INTERMEDIATE TECHNOLOGY DEVELOPMENT GROUP). British based equivalent to VITA, they haven't done much on solar power as yet but seem likely to in the near future.

SOLAR ENERGY LABORATORY, C.N.R.S., PYRENEES, FRANCE. DIRECTOR: PROFESSOR F. TROMBE. Built solar furnace in the Pyrenees, and the Trombe/Michell solar wall houses.

VITA (VOLUNTEERS FOR INTERNATIONAL TECHNICAL ASSISTANCE), 3706 RHODE ISLAND AVE., MT. RAINIER,

MARYLAND 20822, USA. Organisation concerned with low technology applications in the Third World. They have publications and plans for solar collectors, methane plants, water power, and various types of windmills. List of plans available.

TELKES, Maria 'LOW COST SOLAR HEATED HOUSE' Heating and Ventilating Aug. 1950, 47:72. 'REVIEW OF SOLAR HOUSE HEATING' Heating and Ventilating 1949, 46:68 (Sept. 1). 'SOLAR HOUSE HEATING, A PROBLEM OF STORAGE' Heating and Ventilating 1947, 44:68.

THOMASON, Harry E. 'HOUSE WITH SUNSHINE IN THE BASEMENT' Popular Mechanics Feb. 1965 pp 89—92. 'SOLAR HOUSE MODELS' Edmund Scientific, 100 Edscorp Bldg., Barrington, N.J. 08077, 1965. Describes Thomason's solar house in Washington and other designs.

TOILETS, SELECTED LIST OF :-

> CLIVUS. ANDSTOR INTERNATIONAL, Virebergsvagen 7, Box 1023, S-171 21 Solna 1, Sweden.

> HUMUS TOILET. SANITATION AG, Baarerstrasse 59, 6300 Zug. Switzerland. he Humus-toilet is very similar to the Mull-Toa Sanitation AG's publication 'What happens inside the Humus Toilet'. Most interesting.

> KERN COMPOST PRIVY. THE OWNER BUILT HOME, Ken Kern Drafting 1972 pp 68—71

> MULLBANK. INVENTOR AB, Prastgatan 42, 831 00 Ostersund, Sweden.

> MULL-TOA HANS-KR NIELSEN, Sorkadalsveien 22, Oslo 3, Norway.

> MULTRUM. SCAN-PLAN, 3 Sankt Kjelds Gade, DK-2100 Copenhagen, Denmark.

TROMBE, F.A. LE PHAT VINH & MME. LE PHAT VINH 'ETUDE SUR LE CHAUFFAGE DES HABITATIONS PAR UTILISATION DU RAYONNEMENT SOLAIRE' Revue Generale de Thermique Vol. IV, No. 48 Dec. 1965 (in French). Report on and description of construction and operation of the CNRS/Trombe solar wall heated houses in the Pyrenees. Good graphic presentation of the theoretical analysis of insolation values at different latitudes on a range of orientations, etc.

UNESCO 0A54. SYMPOSIUM ON SOLAR ENERGY & WINDPOWER IN ARID ZONES. New Delhi Symposium. UNESCO, Paris 1956.

UNESCO. THE SUN IN THE SERVICE OF MAN. Paris 1973. Large number of papers covering the following fields: space and water heating, large and small scale power production, photovoltaics, solar stills, the sun and plant growth. Bioclimatology. (Summaries of papers releyant to building are in Moorcraft's article, Architectural Design, Oct. 73. op. cit.)

UN PUBLICATIONS. 1961 ROME CONFERENCE ON NEW SOURCES OF ENERGY: PROCEEDINGS OF: UN Publications New York 1964.

Vol. 1 General Sessions. Wide coverage of energy needs, locally available sources of energy and energy storage.

Vol. 4 Solar Energy I for Mechanical Power and Electrical Production. Pistons, turbines, and conversion to electricity.

Vol. 5 Solar Energy II for Heating. Covers most aspects of solar heating.

Vol. 6 Solar Energy III for Cooling. Distillation and furnaces.
Vol. 7 Wind Power

VINCE, J.T. DISCOVERING WATERMILLS. Shire 1970. Very interesting little pocket size book on traditional watermills in Britain, contains useful gazeteer if you wish to go watermill exploring.

VINCZE, Stephen 'A HIGH-SPEED CYLINDRICAL SOLAR WATER HEATER' Solar Energy Journal November 71, 13(3) 339—344. Description of novel type of solar water heater.

VITA. VILLAGE TECHNOLOGY HANDBOOK (AA) (Central Poly). Various bits and pieces about water power and mechanical transmission, bamboo water wheels, flow measurement and hydraulic rams.

WAGNER, E.G. & LANOIX, J.N. WATER SUPPLY FOR RURAL AREAS AND SMALL COMMUNITIES (mono series No. 42) S6 75 World Health **Organisation 1959. Mainly about water supply and filtration, but some** useful info. on dams and water flow measurement.

WATER TURBINE MANUFACTURERS. All turbines are rather expensive, so they tend to be suitable for rich eco freaks only.

ARMSTRONG EVANS, CHAGFORD, DEVON, ENGLAND. Manufacture small water turbines.

GILBERT GILKES & GORDON LTD., CANAL IRON WORKS, KENDAL, WESTMORELAND, ENGLAND LA9 7BZ. Manufacture small turbines, but require specific information. Low head units are expensive and cost of installation becomes more viable the higher the head.

LEFFEL & CO., JAMES, SPRINGFIELD, OHIO, USA. Small Samson turbines available from 3 to 29 horsepower and Hoppes Hydro electric units of 1 to 10kw. capacity.

NEYRPIC, DIVISION DE LA SOCIETE ALSTHOM, BP 75 CENTRE DE TRI-GRENOBLE, CADEX, FRANCE. Mainly large water turbines but may produce some small scale units.

OFFICINE BUEHLER, TAVERNE, CANTON TICINQ, SWITZER-LAND. They are in the small water turbine field and manufacture all types apart from Michell. Workmanship reputed to be top quality.

OSSBERGER TURBENFABRIK, 8832 WEISSENBURG, BAVARIA, W. GERMANY. Exclusive manufacturers of the Michell (or Banki) turbine in sizes ranging from 1 to 1,000 horse power. Have impressive record of installations, many of which in Third World countries. Are very responsive to enquiries and will provide info. in English.

HYDRAULIC RAM MANUFACTURERS

CE CO CO CHUO BOEKI GOSHI KAISHA. PO BOX 8, IBARAKI CITY, OSAKA, JAPAN.

RIFE HYDRAULIC ENGINE CO., BOX 337, MILLBURN, NJ 07041, USA.

WHOLE MOTHER EARTH WATERWORKS, THE, C/O. EDWARD BARBERIE, BOX 104, GREEN SPRING, W. VA. 26722, USA. Will design and build a unit for $17.00 if specifications and requirements sent.

WATSON-MUNRO. REPORT OF COMMITTEE ON SOLAR ENERGY RESEARCH IN AUSTRALIA. Australia Academy of Sciences, Sept. 73.

WIND GENERATORS AND WIND PUMPS, Manufacturers of.

Aerowatt S.A. 37 rue Chanzy, 75-Paris 11ᵉ, France. 30w to 4kw machines.

Carl, William, Plant 30/Dept 300, Grumman Aircraft Corporation, Bethpage, L.T., New York, USA. Manufacturing sail-wing windmills under licence.

Dempster (Annu Oiled Windmills) Industries Inc., PO Box 848, Beatrice, Nebraska 68310, USA. Multivane windpumps.

Dunelite Wind Driven Generators, Division of Pye Ind., 21 Frome St., Adelaide 5000, Australia. 1/2kw machines.

Elektro GmbH. Winterthur (Schweiz), St. Gallerstrasse 27. 0.05 to 6kw machines.

Eoliennes Enag., Electro Mechanique, Route du Ponte L'Abbe, 29 S-Quimper (Finisters), France. 0.18 to 2kw machines.

Godwin Ltd., H.J., Quenington, Glos. GL7 5BX, England, manufacture "Hercules" multivane windpumps, also hand pumps.

Lubing, Maschinenfabrik, Ludwig Bening 2847 Barnstorf, PO Box 171, W. Germany. Wind generators and windpumps.

Metters Building Products (SA) Pty, Ltd., Box 2047, GPO Adelaide, S. Australia 5001. Various sizes of Multivane windpumps.

Quirks Victory Light Co., 38 Fairweather St., Belle Vue Hill, NSW, Australia. Same machines as Dunlite, but more expensive.

Southern Cross Engine and Windmill Co., Pty., Ltd., 1 Grand Avenue, Granville, Sydney, NSW, 2142, Australia. Multivane windpumps.

Winco, Division of Dyna Technology Inc., Sioux City, Iowa 51102, USA. Produce "Wincharger" 200W machine.

A longer list of windmill manufacturers is available from the Technical Research Division, University of Cambridge CB2 1PX, (Technical Bulletin No. 1, by Gerry Smith.)

USEFUL ORGANISATIONS WORKING IN THE FIELD

BRACE RESEARCH INSTITUTE, MacDonald College, McGill University, Ste. Anne de Bellevue 800, Quebec, Canada. They have 60 or so publications on solar energy, water purification, and wind power. Publications list available. Very good work.

ELECTRICAL RESEARCH ASSOCIATION, Cleeve Road, Leatherhead, Surrey, England. Have had some 26 years of experience in the field of wind energy, and has 35 publications about their research (some of which are rather expensive). Some good quality research, but mainly to do with large scale machines.

ITDG (INTERMEDIATE TECHNOLOGY DEVELOPMENT GROUP). British based equivalent to VITA, are just beginning to do work on windmills, but not much produced so far.

VITA (VOLUNTEERS FOR INTERNATIONAL TECHNICAL ASSISTANCE), 3706 Rhode Island Ave., Mt. Rainier, Maryland 20822, USA. Organisation concerned with low technology applications in the Third World with publications and plans for solar collectors, methane plants, water power and various types of windmills. List of plans available.

WINDWORKS (Hans Meyer's group), Box 329, Route 3, Mukwonage, WL 53149, USA. Have done some good work with regard to small-scale self-build machines.

YELLOTT, John UTILISATION OF SUN AND SKY RADIATION FOR HEATING & COOLING OF BUILDINGS. Ashrae Journal, December 1973. An excellent summary of the technical problems, written as a preliminary chapter to the solar energy section of the forthcoming ASHRAE HANDBOOK